Band 19 · Oktober 1983 **10**

isotopenpraxis

AKADEMIE-VERLAG

ISSN 0021-1915 Isotopenpraxis, Berlin 19 (1983) 10, 329-364

EVP 9,- M

An unsere Autoren

Die Zeitschrift veröffentlicht Beiträge zu folgenden Themen: Kontrolle und Analyse radioaktiver und stabiler Isotope, Anwendung radioaktiver und stabiler Isotope in allen Gebieten der Forschung und Technik, der Industrie, Landwirtschaft und Medizin, Herstellung radioaktiver und stabiler Isotope sowie von markierten Verbindungen, kernelektronische Geräte für die Isotopenanwendung (Dickenmesser, Dichtemesser u. ä.) Detektoren für isotopentechnische Arbeiten und den Strahlenschutz in Isotopenlaboratorien, Meßtechnik und Strahlenschutz, Dosimetrie in Isotopenlaboratorien, kerntechnische Geräte für die Isotopenanwendung (Beton, Bestrahlungsanlagen u. ä.), Strahlenchemie, Radiochemie, ökonomische und perspektivische Fragen, Fragen der Weiterbildung.

1. Aufnahme von Manuskripten

Zur Veröffentlichung angenommen werden nur druckreife, noch nicht anderweitig publizierte Beiträge in deutscher, englischer oder russischer Sprache zum Themenbereich der Zeitschrift und zwar

Originalarbeiten
Kurze Mitteilungen
Aus der Praxis
Zusammenfassende Berichte
Tagungsberichte
Informationen.

Sind der Beitrag oder Teile daraus bereits veröffentlicht oder ist eine anderweitige Veröffentlichung vorgesehen, so muß dies der Redaktion mit genauer Angabe der Zweitveröffentlichung mitgeteilt werden. Mit der Annahme des Manuskriptes behält sich der Verlag das Recht der Vervielfältigung, Verbreitung und Übersetzung innerhalb der gesetzlichen Schutzfrist vor. Auszugsweiser Nachdruck oder photomechanische u. a. Wiedergabe sind an eine schriftliche Genehmigung des Verlages gebunden. Manuskripte, Zeichnungen und Photos von veröffentlichten Arbeiten werden nur auf den besonderen und bis zum Zeitpunkt der Veröffentlichung mitgeteilten Wunsch des Autors zurückgesandt.

Bei der unverlangten Einreichung von Manuskripten wird empfohlen, eine Befürwortung eines der Herausgeber oder Redaktionsausschußmitglieder beizufügen.

Manuskripte sind einzureichen bei der Redaktion, 1080 Berlin, Leipziger Str. 3–4

2. Form der Manuskripte

Die Autoren werden gebeten:

2 Manuskripte, eines davon in Original-Maschinenschrift, einseitig und mit 2zeiligem Abstand einzureichen; gut leserliche Photokopien sind zulässig; auf dem 1. Blatt (unter Weglassung der Bezeichnung)

den Titel der Arbeit
Vorname-Initialen und Namen sämtlicher Autoren
gegebenenfalls die Institution, in der sie tätig sind
eine kurze Zusammenfassung der Arbeit von höchstens 15 Zeilen
Keywords in alphabetischer Reihenfolge und
als Fußnote die Anschrift der Institution anzugeben;

wenn möglich, auf gesonderten Blättern Namen, Titel der Arbeit und Zusammenfassung in den beiden anderen obengenannten Sprachen 2fach beizufügen;

Literaturhinweise (fortlaufend numeriert), Fußnoten, Tabellen, Abbildungen (in 2–3facher Veröffentlichungsgröße), Abbildungsunterschriften (getrennt von den Abbildungen mit Zeichenerklärungen, Quellenangaben, falls erforderlich, und gegebenenfalls mit einem Vermerk für besondere Wünsche bezüglich der Abbildungsgröße), private und Dienstanschrift der Verfasser mit akademischen Titeln und ausgeschriebenen Vornamen auf besonderen Blättern ans Ende des Manuskriptes zu stellen;

Wünsche für den Kolumnentitel (Seitenüberschrift) bekanntzugeben;

Absätze durch Einzüge von mindestens 3 Anschlägen kenntlich zu machen;

die Maßeinheiten des SI-Systems ("Systeme International") zu verwenden;

Korrekturen im Manuskript unmittelbar an der zu korrigierenden Stelle anzubringen;

Formeln deutlich lesbar und möglichst groß einzutragen;

Farbabbildungen und nachträgliche Änderungen oder Ergänzungen in Abbildungen zu vermeiden;

hervorzuhebende Textstellen zu unterstreichen, und zwar halbfetten Satz mit einer geraden schwarzen Linie, kursiven Satz (Schrägschrift) mit einer blauen Wellenlinie (alle Eigennamen). Sperrungen gestrichelt;

besondere Buchstaben durch farbige Unterstreichung zu kennzeichnen, und zwar Formelzeichen in

Griechisch-Kursiv (α, β, ...) rot,
Kursiv (A, B, ...) blau
halbfetter Kursiv ($\boldsymbol{A}$, $\boldsymbol{B}$, ...) 2 Linien blau (Vektoren),
halbfetter Grotesk ($\mathbf{A}$, $\mathbf{B}$, ...) 2 Linien violett (Tensoren);

bei Literaturangaben auf Vollständigkeit zu achten. In nachstehender Reihenfolge müssen angegeben sein:

a) bei *Zeitschriften*: Vorname-Initialen und Namen sämtlicher Autoren, Titel der Zeitschrift (nach „Periodica Chimica" oder möglichst wenig gekürzt), Bandnummer (unterstrichen), Jahreszahl (in Klammern), Heftnummer (nur bei Zeitschriften, die jedes Heft mit S. 1 beginnen) sowie die Seitenzahl, auf der die zitierte Arbeit beginnt.

b) Bei *Büchern*: Initialen der Vornamen und Name des Autors (Herausgeber werden durch den Zusatz „ed." gekennzeichnet), voll ausgeschriebener Titel des Buches in der Originalsprache, evtl. Bandnummer, voll ausgeschriebener Name des Verlages, Nummer der Auflage (außer der ersten), Erscheinungsort(e), Erscheinungsjahr (hier nicht in Klammern setzen!) und evtl. Kapitelnummer, Seitenzahl u. ä.

c) bei *Sammelbänden, Tagungsberichten*: Es soll zusätzlich der Titel des zitierten Beitrags angegeben werden. Die Seitenzahl muß unbedingt angegeben werden. Aus dem Zitat muß hervorgehen, ob es sich um einen Tagungssammelband oder um einen noch nicht veröffentlichten Vortrag handelt. Zwischen Tagungsort und Erscheinungsort des entsprechenden Tagungsberichts ist zu unterscheiden.

3. Korrekturen

Der Autor erhält zwei Umbruchabzüge, von denen einer möglichst schnell korrigiert an die Redaktion zurückgereicht werden soll. Auf Wunsch des Autors kann die Korrektur auch von der Redaktion gelesen werden.

Durch nachträgliche Manuskriptänderungen verursachte Korrekturkosten gehen, sofern sie 10% der Satzkosten übersteigen, zu Lasten des Autors.

Der Autor erhält 50 Sonderdrucke kostenlos. Weitere Sonderdrucke können bis zur Rückgabe der Korrekturabzüge auf Kosten des Autors bei der Redaktion bestellt werden.

Bezugsmöglichkeiten

Bestellungen sind zu richten

- **in der DDR** an den Postzeitungsvertrieb unter Angabe der Kundennummer des Bestellenden oder an den AKADEMIE-VERLAG, DDR-1086 Berlin, Leipziger Straße 3–4, PF-Nr. 1233;
- **im sozialistischen Ausland** an eine Buchhandlung für fremdsprachige Literatur oder an den zuständigen Postzeitungsvertrieb;
- **in der BRD und Berlin(West)** an eine Buchhandlung oder an die Auslieferungsstelle KUNST UND WISSEN, Erich Bieber OHG, Wilhelmstraße 4–6, D-7000 Stuttgart 1;
- **in den übrigen westeuropäischen Ländern** an eine Buchhandlung oder an die Auslieferungsstelle KUNST UND WISSEN, Erich Bieber GmbH, Dufourstraße 51, CH-8008 Zürich;
- **im übrigen Ausland** an den Internationalen Buch- und Zeitschriftenhandel; den Buchexport, Volkseigener Außenhandelsbetrieb der Deutschen Demokratischen Republik, DDR-7010 Leipzig, Postfach 160; oder an den AKADEMIE-VERLAG, DDR-1086 Berlin, Leipziger Straße 3–4.

Zeitschrift „Isotopenpraxis"

Herausgeber: Im Auftrag des Akademie-Verlages herausgegeben von Prof. Dr. Justus Mühlenpfordt
Verlag: Akademie-Verlag, DDR-1086 Berlin, Leipziger Straße 3–4, PF-Nr. 1233; Fernruf: 2236229; Telex-Nr.: 114420;
Bank: Staatsbank der DDR, Berlin, Kto.-Nr.: 6836-26-20712
Verantwortlich für den Inhalt des Heftes: Prof. Dr. H. Hübner, Leipzig
Anschrift der Redaktion (Fernruf Frau H. Schier: 6351177) der Zeitschrift Isotopenpraxis: Akademie-Verlag, DDR-1086 Berlin, Leipziger Str. 3–4, PF-Nr. 1233
Veröffentlicht unter der Lizenznummer 1527 des Presseamtes beim Vorsitzenden des Ministerrates der Deutschen Demokratischen Republik
Gesamtherstellung: VEB Druckerei „Thomas Müntzer", DDR-5820 Bad Langensalza
Erscheinungsweise: Die Zeitschrift „Isotopenpraxis" erscheint monatlich. Die 12 Hefte eines Jahrganges bilden einen Band. Bezugspreis je Band 180,– M zuzüglich Versandspesen (Preis für die DDR 108,– M). Preis je Heft 15,– (Preis für die DDR 9,– M)
Bestellnummer dieses Heftes: 1079/19/10

AN (EDV) 12831

ISOTOPENPRAXIS

Band 19 (1983) Heft 10, S. 329–364

Redaktion und Verlag: Akademie-Verlag, Redaktion Isotopenpraxis, 1086 Berlin, Leipziger Straße 3–4

Inhaltsverzeichnis

Originalarbeiten

R. P. Gardner, K. Verghese

Anwendung von Monte-Carlo-Methoden bei der Kalibrierung und Optimierung der Auslegung von Strahlungsmeßgeräten und Strahlungsanalysatoren (*in Englisch*)

Isotopenpraxis 19 (1983) H. 10, S. 329 bis 335
1 Abb., 1 Tab., 51 Lit.

Monte-Carlo-Methoden, eingesetzt für die Kalibrierung und optimale Konstruktion von Strahlungsmeßgeräten und Strahlungsanalysatoren, werden beschrieben. Die Vorteile der Anwendung spezifischer Monte-Carlo-Methoden für bestimmte Meßgeräte oder Analysatoren gegenüber der Anwendung vorhandener herkömmlicher Monte-Carlo-Codes werden diskutiert. Der umfassende Einsatz von Varianz-Reduzierungsverfahren, die auf den physikalischen Prinzipien und Eigenschaften des interessierenden Systems beruhen, erfolgt bei der Auslegung dieser Monte-Carlo-Modelle.

W. Birkholz, M. Steinert

Auswertung von Autoradiogrammen mit Hilfe eines Mikrorechners (*in Englisch*)

Isotopenpraxis 19 (1983) H. 10, S. 335 bis 338
7 Abb., – Tab., 7 Lit.

Das System AURAS zur Auswertung von Autoradiogrammen besteht aus einem Digitalphotometer, dem Densitron, einem Fernseh-Scanner zur Digitalisierung der Aufnahmen und einem Mikroprozessor zur Bildauswertung und Datenspeicherung. Das Digitalphotometer erlaubt nicht nur eine sehr genaue Auswertung, sondern ist auch sehr zeitsparend. Es wurde ferner ein Programmsystem MARAUS entwickelt, das auch im Dialogregime arbeiten kann.
Einzelheiten über die Auswertung von Dichtehistogrammen und Flächen mit unterschiedlichen Schwärzungen werden mitgeteilt und diskutiert.

L. Petryka, K. Przewłocki

Radiotracer-Untersuchungen bei der Aufbereitung von Kupfererz in industriellen Flotationsprozessen (*in Englisch*)

Isotopenpraxis 19 (1983) H. 10, S. 339 bis 341
3 Abb., 2 Tab., 3 Lit.

Die Optimierung der Aufbereitung von Kupfererz erfordert die Identifizierung des gesamten Prozesses, bei dem die Flotation die Hauptoperation darstellt. Natürliches polymetallisches Erz, in das ^{64}Cu-Aktivität induziert worden war, diente als Tracer. Gegenstand der detaillierten Messungen war eine handelsübliche 14-Kammer-Flotationsanlage. Für alle Teile der Flotationsanlage wurden die mittlere Verweilzeit für die verschiedenen Korngrößenfraktionen des natürlichen Kupfersulfids sowie die Wirksamkeit der Flotation ermittelt.

H. Langguth, H.-U. Försterling, J. Schmidt

Gamma-Radiolyse von Dimethylglyoxim

Isotopenpraxis 19 (1983) H. 10, S. 342 bis 344
5 Abb., 1 Tab., 17 Lit.

Die Strahlenresistenz von reinem und an Aktivkohle sorbiertem Dimethylglyoxim (DMG), das für die selektive Abtrennung von Pd aus Abfallösungen der Kernbrennstoffwiederaufbereitung geeignet ist, wurde untersucht. Die Bestrahlung erfolgte bei 20–40 °C mit einer

Contents

Articles

R. P. Gardner, K. Verghese

Use of Monte Carlo Methods in the Calibration and Design Optimization of Radiation Gauges and Analyzers (*in Englisch*)

Isotopenpraxis 19 (1983) No. 10, p. 329 to 335
1 Fig., 1 Tab., 51 Ref.

Monte Carlo Methods are described for use in the calibration and optimum design of radiation gauges and analyzers. The advantages of the use of specific Monte Carlo models designed for a particular gauge or analyzer are discussed relative to the use of existing general purpose Monte Carlo codes. The extensive use of variance reduction techniques based on fundamental physical principles and properties of the particular system of interest is stressed in the design of these Monte Carlo models.

W. Birkholz, M. Steinert

Evaluation of Autoradiograms Using a Microcomputer (*in English*)

Isotopenpraxis 19 (1983) No. 10, p. 335 to 338
7 Fig., – Tab., 7 Ref.

An equipment AURAS for evaluation of autoradiograms was developed. It consists of a digitalphotometer and the densitron system (television scanner) for digitalisation of pictures and a microprocessor for picture processing and storing of datas.
The digitalphotometer permits a precise but timeconsuming scanning. The densitron system works quickly with little density classes and pseudocolouring of pictures.
For the evaluation of autoradiograms with the microcomputer a programsystem MARAUS exists. It works in a dialogeous regime and has following functions:
data transfer between the scanners and the computer; estimation of density histograms and areas of different density classes in the picture; calculation of average of densities; printing of density profiles or classifite pictures; transfer of data to external medias such as punched tape or magnetic cassette for picture processing on bigger computers. The possibilities of using the equipment for evaluation of autoradiograms are demonstrated.

L. Petryka, K. Przewłocki

Radiotracer Investigations of Benefication Copper Ore in the Industrial Flotation Process (*in English*)

Isotopenpraxis 19 (1983) No. 10, p. 339 to 341
3 Fig., 2 Tab., 3 Ref.

Optimization of copper ore benefication requires identification of the whole process, in which flotation is the main operation. Natural polymetalic ore in which activity of ^{64}Cu was induced served like a tracer. Subject of detailed measurements was a 14 chamber commercial flotation. machine. As a result mean residence time for different granulometric fractions of natural copper sulphide was determined for all parts of flotation machine, as well as the efficiency of flotation.

H. Langguth, H.-U. Foersterling, J. Schmidt

Gamma Radiolysis of Dimethylglyoxime

Isotopenpraxis 19 (1983) No. 10, p. 342 to 344
5 Fig., 1 Tab., 17 Ref.

Because dimethylglyoxime (DMG) is suitable for selective separation of palladium from waste solutions of fission products of nuclear power stations the radiation resistance of DMG pure and sorbed on activated charcoal was studied. Irradiation was carried out at 20–40 °C using a

^{60}Co-γ-Bestrahlungsanlage mit Dosisleistungen von 0,6 bis 1,7 Gy/s in einem Dosisbereich von 2 kGy bis 5 MGy. Die Radiolyseprodukte wurden mit verschiedenen Methoden analysiert und die G-Werte bestimmt. Es wurde gezeigt, daß DMG bis zu Dosen von etwa 100 kGy relativ strahlenresistent ist. Bei höheren Dosen werden die reaktiven NOH-Gruppen des GMD teilweise zerstört, so daß sie nicht mit Pd reagieren können. Im Verfahren zur Abtrennung von Pd aus radioaktiven Abfallösungen werden diese Dosen jedoch nicht erreicht. Auf eine unzureichende Strahlenresistenz von DMG zurückzuführende Komplikationen können deshalb ausgeschlossen werden.

^{60}Co-γ-source with a dose rate of 0,6 – 1,7 Gy/s in the dose range 2 kGy to 5 MGy. The radiolysis products were analysed by different methods and the G-values determined. It has been shown, that DMG is relatively radiation resistant up to doses of about 100 kGy. At higher doses the reactive NOH-groups of DMG are partially destroyed and cannot react with palladium. However, in the separation process of palladium from radioactive waste solutions these doses will not be reached so that complications due to insufficient radiation resistance of DMG can be excluded.

M. Gläser, B. Herold

„Gammadefektoskopie" an Kartoffeln mit ^{241}Am

Isotopenpraxis 19 (1983) H. 10, S. 345 bis 350
9 Abb., 3 Tab., 6 Lit.

Die zunehmende Industrialisierung der landwirtschaftlichen Produktion bei höheren Qualitätsanforderungen verlangt objektive und hochempfindliche Kontrollverfahren.
Zur zerstörungsfreien Ermittlung innerer Qualitätsmängel an Früchten, z. B. Kartoffelknollen, können lichtoptische Verfahren wegen der starken Streuung nicht umfassend eingesetzt werden. Deshalb wurde unter Einsatz weicher Gammastrahlung (^{241}Am) versucht, „trockene" Hohlstellen, insbesondere bedingt durch Hohlherzigkeit, in Kartoffelknollen zu erkennen. Es konnte gezeigt werden, daß bei einem Erkennungsfehler von etwa 13 % diese Mängel (ab ca. 5 mm Durchmesser) zu erkennen sind, auch wenn die Knollen nur in einer durch die Knollenmitte verlaufenden Ebene durchstrahlt werden.

M. Glaeser, B. Herold

"Gammadefectoscopy" on Potatoes with ^{241}Am

Isotopenpraxis 19 (1983) No. 10, p. 345 to 350
9 Fig., 3 Tab., 6 Ref.

The mechanisation of modern farming is characterized by introduction of more objective and high sensitive monitoring technique. For instance this is important for the evaluation of quality properties of agricultural products such as fruits and vegetables. The nondestructive determination of diseases on fruits can be done with light measuring technique (color), but it is impossible to discover internal defects such as "hollow heart" in potatoes. For this purpose was used soft gamma radiation from radionuclid ^{241}Am. It could be showed to discover such internal defects greater than 5 mm in diameter with an error of perception of 13 percent, even when the radiation was directed through a little area perpendicular to the lengthwise moving potatoes.

Kurze Originalmitteilungen

G. Bernhard

Untersuchungen zum Adsorptions-Desorptions-Verhalten von $^{99}MoO_4^{--}$ an Al_2O_3 in Abhängigkeit von der Temperatur

Isotopenpraxis 19 (1983) H. 10, S. 351 bis 352
3 Abb., – Tab., 3 Lit.

Letters of the Editor

G. Bernhard

Studies on the Adsorption and Desorption Behaviour of $^{99}MoO_4^{--}$ on Al_2O_3 as a Function of Temperature

Isotopenpraxis 19 (1983) No. 10, p. 351 to 352
3 Fig., – Tab., 3 Ref.

Informationen

Isotopenpraxis 19 (1983) H. 10, S. 353 bis 358

Information

Isotopenpraxis 19 (1983) No. 10, p. 353 to 358

Buchbesprechungen

Isotopenpraxis 19 (1983) H. 10, S. 358 bis 364

New Books

Isotopenpraxis 19 (1983) No. 10, p. 358 to 364

Isotopenpraxis 19 (1983) 10, S. 329–335

Use of Monte Carlo Methods in the Calibration and Design Optimization of Radiation Gauges and Analyzers*)

R. P. Gardner, K. Verghese

(Center for Engineering Applications of Radioisotopes,
Department of Nuclear Engineering, North Carolina State University)¹)

Monte Carlo Methods are described for use in the calibration and optimum design of radiation gauges and analyzers. The advantages of the use of specific Monte Carlo models designed for a particular gauge or analyzer are discussed relative to the use of existing general purpose Monte Carlo codes. The extensive use of variance reduction techniques based on fundamental physical principles and properties of the particular system of interest is stressed in the design of these Monte Carlo models.

Keywords

calibration; calibration standards; computer-aided design; measuring instruments; Monte Carlo method; radiation detection

1. Introduction

This paper primarily describes the experience of the authors in using Monte Carlo methods in the calibration and design of radiation gauges and analyzers. It is not intended as a general review of this subject and, therefore, may not include many references that are generally pertinent to this subject.

The paper first gives a brief review of probabilistic Monte Carlo Methods including descriptions of the "crude" or "hit-or-miss" method at one extreme and the "complete variance reduction" method at the other. The various variance reduction methods are categorized and discussed. Then our experiences in using existing general purpose Monte Carlo codes are described. Following this our work on developing specific Monte Carlo models is described; first for radiation gauges and then for radiation analyzers. Here emphasis is placed on the problems encountered in each case. Finally some general conclusions are drawn and discussed relating to future uses of Monte Carlo methods for the calibration and optimum design of radiation gauges and analyzers.

It is our opinion that the use of Monte Carlo methods should become more common, particularly in fields like radioisotope and radiation applications which have traditionally been oriented primarily to development by experimental methods. In a sense probabilistic Monte Carlo is an experiment. Therefore, an experimentalist should have a natural advantage in adapting this technique to add a mathematical modelling capability to his or her set of capabilities.

2. A Brief Review of Monte Carlo Methods

The probabilistic Monte Carlo method essentially consists of following individual histories of the radiations or particles of interest by randomly selecting the possible angles, path lengths between interactions, and energies from the appropriate probability distribution functions (pdf's) of the pertinent interactions with matter [1–7]. The yield (number of successes divided by the total number of histories) of a large number of these histories is an estimate of the probability of the phenomenon of interest.

When this method is applied in the simplest possible way without restriction of the variables that control the paths that are chosen, then it is known as "crude" or "hit-or-miss" Monte Carlo. This approach has the advantages that the yields obtained contain no bias or error and the model development requires the least amount of skill and ingenuity. However, the disadvantage is that very large numbers of histories are usually required to attain a reasonable accuracy. For many applications and most of those pertaining to the present subjects, this number is so large that it precludes the use of this method entirely.

Fortunately, one can usually use variance reduction techniques to reduce the number of histories required to attain a given accuracy. A variance reduction technique is defined here as any modification of the crude or hit-or-miss Monte Carlo method that reduces the variance (or standard deviation) of the calculated yield per unit history. By judicious use of variance reduction methods it is often possible to achieve "complete variance reduction" so that every history results in a success (weighted by an appropriate fractional value). This represents the other exteme compared to crude or hit-or-miss Monte Carlo.

There are different types of variance reduction methods. For the present purposes they are categorized as: (1) exact analytical replacements, (2) variable range restrictions with appropriate weighting, and (3) purely mathematical methods. The first of these consists of solving part of the problem by exact (deterministic analytical or experimental) methods when this can be done and using this exact solution in place of the Monte Carlo method for that part of the problem. This should always be done when it is possible since, by definition, it is an exact replaement and, therefore, introduces no bias or error. An example of this type of variance reduction method is the use of detector

*) Presented at the 2nd Working Meeting "Radioisotope Application and Radiation Processing in Industry", 28th September to 1st October 1982, Leipzig
¹) Address: Raleigh, North Carolina 27650, USA

efficiencies. The Monte Carlo method can be used up to point that the radiation or particle involved is incident on the detector of interest. At this point one finishes the calculation by simply multiplying the answer by the appropriate value of detector efficiency.

The second type of variance reduction consists of restricting the path of the radiation or particle to within certain ranges of the pertinent variables that are capable of yielding a success. When such restrictions are made one formally accounts for them by changing (usually reducing) the weight for that history in an appropriate manner. A general rule for determining whether or not a variance reduction of this type is valid and will not introduce bias is to ascertain whether or not any (or a significant fraction) of the histories can be successful from the value of the variable (position) in question. This type of variance reduction can often be very significant. It has the advantage that it is a "common sense" or "natural" type of variance reduction method for a person that is knowledgeable about the process or phenomenon in question. That is, such a person will know automatically whether the variance reduction method in question is valid or not. An example of this type of variance reduction is the limiting of the reflection or backscatter of radiation to a thickness beyond which it is known from experimental evidence that a negligible contribution is made.

The third and final type of variance reduction is called the "purely" methematical to emphasize that this type usually has no direct physical basis. That is, one loses the relationship between the variance reduction method being used and the physics of the process or phenomenon. The specific variance reduction methods that sometimes (but not always) fit into this category are: (1) splitting, (2) stratified sampling, (3) Russian roulette, and (4) variable transforms such as the exponential or linear transforms for distances between radiation interactions. One should be very careful in applying this type of variance reduction because of the danger of introducing a bias into the calculation.

3. Use of Existing General Purpose Monte Carlo Computer Codes

Our first experience with Monte Carlo methods was also our first experience with the use of existing general purpose Monte Carlo codes. To simulate neutron moderation surface-type soil moisture content gauges the 05R Computer Code was used [8–11]. This code was originally developed as a relatively limited reactor physics (neutron transport) code for the IBM-704 computer by *R. R. Coveyou* in 1958. Subsequently it evolved (8, 9) into a much more versatile and general neutron transport code.

The 05R code contained a very detailed representation of neutron cross sections. There were 40 energy "supergroups" with boundaries a factor of two apart from 0.000070 eV to 77.13 MeV. Each supergroup was then divided into n subgroups of equal energy where n is $2m$ and m could be an integer from one to nine. To avoid using the very large amount of computer memory necessary to store all this cross section data the computer program would store only the data in one supergroup at a time and would process batches of neutrons through this supergroup in a parallel fashion.

The 05R code contained a very general geometry subroutine which permitted the treatment of very complicated configurations. The interface surfaces could be either plane or quadratic, could be oriented in any way, and could intersect in arbitrary fashion. The complexity of the geometry to be considered was literally limited only by the amount of computer memory that one was willing to commit to this aspect of the problem.

The major built-in variance reduction method in the 05R code was that neutron absorption was not permitted. An additional variance reduction was available in that arbitrary weights could be assigned for each initial neutron of a given speed, set of spatial coordinates, and set of direction cosines. This feature allowed one to use various stratified or importance sampling schemes as specified by the user. The path of each neutron was followed in the usual way until it reached thermal (0.025 eV) energy. After this the neutron was given a simple one velocity group treatment in which in all scatters the neutron neither gained nor lost energy and the scattering was considered to be isotropic in the laboratory system.

Even with this generality a number of modifications were necessary to apply the 05R code to the neutron moderation gauge problem. The 05R code had no provisions for inelastic scattering or multiple energy sources. Modifications were made to accomodate these two things. The neutron moderation surface-type soil moisture content gauge typically has a yield of only one in a thousand or less for the number of neutrons detected per neutron emitted by the source. This means that about one million histories would have to be followed to obtain about one thousand counts in the detector. Fortunately, the symmetry of the problem provided the basis for a significant variance reduction scheme. The neutron moderation gauge problem was cylindrically symmetric about a line through the source perpendicular to the sample surface. That is, the problem of interest consisted of a point source of neutrons located on the surface of a sample which could be considered to be an infinite homogeneous half space. Therefore, the number of neutrons emitted at any angle and energy at a particular radial distance from the source on the sample surface could be considered to be independent of the azimuthal angle. To take advantage of this symmetry the problem was split into two parts. In the first part neutron histories emitted at the sample surface as a function of exit angle, energy, and radial distance from the source were generated and saved. In the second part these stored histories were used with a detailed description of the detector of interest (a BF_3 proportional counter) to estimate the number of detected neutrons. The obvious variance reduction technique of considering only those neutrons emitted by the source into the sample was also used. These two variance reduction methods allowed the use of 9000 histories (of which about 500 were emitted at the sample surface) to provide predicted counting rates for various samples to a relative standard deviation of 4 to 5 percent.

While the 05R Computer Code was successfully adapted to the problem of interest, this was only accomplished at considerable cost. Most of the time spent by the graduate student working on this project (*W. J. Lippold*) was on strictly computer systems related problems. A number of unexpected problems were encountered in attempting to run this code on the IBM 360/75 computer available to us at the Triangle Universities Computation Center at the Research Triangle Park in North Carolina. Typical of these problems was when unknown computer incompatibilities prevented the reading of the binary master cross section tape which was produced by Oak Ridge National Laboratories on their IBM 360/75 Computer. This caused a number of modifications to be made in the main 05R code to accomodate the use of an alternate master cross section tape written in card image form. Other problems were encountered in the amount of computer memory required and in minor differences in the FORTRAN I anguage used. In addition to this the computer time required was extensive. The computer run costs for the entire project which finally yielded predicted counting rates for 19 samples was about $ 10,000.

Our second experience with existing general purpose Monte Carlo codes was in the use of OGRE (and OGRE-G) to simulate the gamma-ray backscatter surface-type soil density gauges [12—15]. This code was originally developed by *Trubey, Penny,* and *Emmett* [12] to study the transmission of gamma rays through laminated slabs. It was subsequently modified by *Trubey* and *Emmett* [13] to utilize the general geometry subroutine written by *Irving, Freestone,* and *Kam* [9] for the 05R Computer Code. This code calculated the absorbed dose rate at up to 10 locations within the chosen geometry. The structure of the cross section data used in this code was identical to that used in the 05R code. For the gamma-ray gauge simulation 11 energy supergroups were used, each with 64 subgroups of equal energy width. The resulting energy range covered was from 16 keV to 32 MeV.

The OGRE-G code is based on the use of the last-flight estimator approach. At each normally selected interaction point within a history the gamma ray is forced to undergo scattering interactions that result in successful detection events at each of the 10 detector locations. Then the gamma ray is allowed to undergo any arbitrary (randomly selected) scattering interaction and a new energy and path length to the next interaction point is selected. This procedure is repeated until the resulting history gamma ray energy is less than the user specified threshold value (a value of 10 keV was used in the present study).

Two major modifications had to be made within the OGRE-G code for simulating the gamma-ray backscatter gauge. The first of these was to add the capability of producing scattered photon spectra. Ten energy increments were established to cover the range from 10 to 662 keV (^{137}Cs was the source of interest). The second modification involved changing the calculation from that of flux to current since it was necessary to know the number of scattered photons that crossed the detector face.

As in the previous case of using the 05R code to simulate the neutron gauge, the use of the OGRE-G code to simulate the gamma-ray gauge involved a problem that was cylindrically symmetric about a line through the point source perpendicular to the sample surface. Therefore, the currents perpendicular to the surface for various gamma-ray energies were first determined as a function of radial distance from the point source. Then a specially designed program was used to predict the counting rate of a right circular cylindrical detector placed at an arbitrary distance with circular face flush with the sample surface using the appropriate scattered gamma-ray currents that had been stored.

One additional important variance reduction technique was also used in this study that was developed by *Gardner, Dunn, McDougall,* and *Lippold* [16]. The basic premise is that the cumulative number of scattered gamma rays emerging from the surface of any given material should depend only on the optical variable dr where d is the material density and r is the radial distance along the surface to the source. Thus, for some other radial distance r' and density d' the cumulative current of scattered gamma rays is identical to that for the same material at the radial distance r and density d provided that

$$d'r' = dr \tag{1}$$

This approach provided a very large savings in the overall simulation program in that the results obtained for a single material density could be generalized to obtain results at any density of interest by this approach.

Even though the OGRE-G Computer Code was also sucessfully applied to the problem of interest, this was again only accomplished at considerable cost. The same types of problems encountered in using the 05R Computer Code were also encountered in the use of the OGRE-G Computer Code. The tape containing gamma-ray cross section data could not be read and several other system specific problems were encountered. Moreover, the graduate student working on this project (*W. C. Hopkins*) spent an inordinate fraction of his time on computer specific systems problems which did not contribute to his understanding of the basic physics pertinent to the problem.

Our third and final experience with the use of existing general purpose computer codes was with the use of the MORSE code to simulate the neutron capture prompt gamma-ray analyzers used for bore-hole logging analysis [17—19]. *Straker* and *Simmons* [18] state: "The very nature (multi-dimensional problems) of the types of problems for which the method is applied makes it difficult to develop a general purpose code with all possible combinations of sources, detectors, etc. The approach that was used in developing MORSE was to provide a basic framework for all the phases and data needed to solve the transport equation and to leave the source and detector descriptions up to the user."

The MORSE code was designed as a multipurpose neutron and gamma-ray Monte Carlo code. It has the ability to treat the transport of either neutrons or gamma rays or to treat a coupled neutron and secondary gamma-ray problem. It is designed for the incorporation of multigroup cross sections that are identical to those used in many other large computer codes such as the ANISN and DOT codes. The multigroup approach allows one to treat either neutrons or gamma rays in the same fashion. In addition, the generation of secondary gamma rays from neutron interactions may be considered as just another group-to-group transfer. The options within the program include the ability to solve either the forward or adjoint problem, the ability to generate albedoes at any material boundary, and the ability to use several types of optional importance sampling approaches.

In using the MORSE code to simulate the neutron capture prompt gamma-ray analyzers for bore-hole logging, two major internal modifications were necessary. The first of these was necessary in order to obtain the discrete gamma-ray energies of interest that are characteristic of the elements in the sample. The existing treatment within the MORSE code yielded only the intensities of multiple energy groups, which was unsuitable for the present application. The other modification was that an updated set of the multigroup cross section data library was required for the neutrons and gamma rays.

The usual additions to the MORSE code relating to the source and detector had to be made. In the present case a ^{252}Cf point source and a NaI(Tl) right circular cylindrical detector were used as in a typical bore hole sonde geometry. This arrangement has cylindrical symmetry about the center line that is coicident with the source and detector major axes.

A post-doctoral fellow from Poland sponsored by the International Atomic Energy Agency (*J. Gyurcsak*) worked on this project. As was the case with both of the graduate students working on the 05R and OGRE-G codes, most of his time was spent on computer specific systems problems and code modifications. The calculations that were finally made for a sample that consisted of sodium chloride (NaCl) dissolved in water, yielded relative standard deviations as high as 60 to 70 percent for the gamma rays of interest for computer run times on the IBM 370/165 computer that were of the order of 12 minutes for a single sample composition.

In all cases it was found that the multigroup cross section approach used in the existing general purpose Monte Carlo codes has the major disadvantages that: (1) one must rely on someone else (usually the code supplier) to generate and provide processed cross section data in a suitable form for use with the computer code of interest and (2) the cross section data is averaged according to some assumed flux distribution (usually a fission spectrum is assumed for neutrons). In practice one may have to wait for some time to obtain the desired processed cross section data and one may have a flux distribution different from that assumed in the calculations.

Another major disadvantage to the use of these general purpose Monte Carlo codes is that formulating real detector responses to the generated flux or current data is often not a simple matter. While it is quite simple to obtain dose rates from fluxes calculated by these codes, actual detectors that do not conform to such idealized responses often cannot be easily modeled. Often one gains a great deal of computational efficiency by including the detector as an integral part of the Monte Carlo model.

4. Specific Monte Carlo Models for Radiation Gauges

With the recent general availability of low cost, reliable, and fast mini- and microcomputers new possibilities exist for use of Monte Carlo models. This includes the laboratory and field use of these smaller computers programmed with Monte Carlo models to provide specific answers to calibration and analysis problems in interactive or near real time calculational modes in locations far removed from the large centrally located computers. This type of application has slightly different programming requirements, particularly in those aspects like the storage of cross section data that require either large amounts of computer memory or the use of other types of off-line storage. A premium is also placed on the development of simple but accurate calculational models for those parts of the Monte Carlo models that are used repetitively and can be treated separately. This often includes the calculation of detector solid angles, efficiencies, and response functions which are discussed in more detail in the following.

Two specific Monte Carlo models are presently being developed for radiation gauges by the authors with the use of these smaller computers in mind. They are for the neutron moderation surface-type moisture content gauges and the gamma-ray backscatter surface-type density gauges that were previously simulated for soil samples with the 05R and OGRE-G codes, respectively.

The use of the Monte Carlo method for simulating radiation gauges is a relatively straightforward application. In most cases if the source involved is a radioisotope it can be considered to be a point and to emit the radiations of interest isotropically. The sample is often either a slab or an infinite half space of uniform density and composition. The detector almost always has the shape of a right circular cylinder. For a given source, sample, and detector with a particular geometrical design a single response (counting rate) is obtained.

In Monte Carlo simulations of this type there are a number of recurrent generic problems to be solved. These often include calculations of the solid angle subtended by and the efficiency of right circular cylindrical detectors. In those cases where variable electronic discrimination for detector responses that are capable of spectrometry is used, the detector response function is another calculation of general interest. The detector response function $R(E, E')$ is defined here as the fraction of the incident radiation of energy E that is detected with resultant pulse-height energy E' to E' plus dE' per unit detector pulse-height energy. It is a probability density function (pdf) in that the integral of $R(E, E')\,dE'$ over all E' is unity.

The solid angle subtended by a detector is useful in those problems in which the attenuation of the radiation from the point of interest to the detector is either negligible or can be taken as constant for the entire detector surface (this latter are large) and when the detection efficiency is either 100 percent for any incident radiation or can be considered a constant for any point of incidence on the detector surface. For example, these conditions are met by a beta-particle source of relatively high energy being detected by an end-window GM tube in air. The authors have developed analytical models for the solid angle subtended from a point at an arbitrary position relative to both a plane circular disc detector or a right circular cylindrical detector [20–22]. These models are calculationally efficient and can be used directly in Monte Carlo models designed for the smaller (and slower) computers.

Detector efficiencies are useful in most all Monte Carlo problems, but are particularly useful when it can be assumed that it is independent of the position of radiation incidence. The authors have developed Monte Carlo models for the detection efficiency of GM tubes and NaI(Tl) crystals for gamma rays [23, 24]. Results from these models could proobably be generalized by empirical fitting methods for inclusion in other Monte Carlo models for the smaller computers.

Detector response functions can also be very useful in Monte Carlo programs, particularly when detector systems capable of spectrometry are being employed. The authors have worked on the development of simple approximate detector response functions for the NaI(Tl) detector [25, 26] and on a relatively elaborate and more accurate one for the Si(Li) detector [27–29]. These models are suitable for inclusion in Monte Carlo models to be used in the smaller computers.

A mentioned previously, one of the major problems associated with the development of specific Monte Carlo models is the preparation, storage, and use of radiation cross section data. The approach adopted by the developers of existing general purpose codes such as the 05R, OGRE-G and MORSE is to form extensive standardized tabular data for every element for many energy groups. This data is then used for specific radiation energies within the Monte Carlo program by choice of the appropriate energy cross section group. This approach has several advantages for the general purpose codes including: generality, the ability to easily update the data, and relatively small access times for the computer. However, the disadvantages for use with specific Monte Carlo models for the smaller computers are that: only a small fraction of the cross section data is usually required for a specific problem, a very large amount of computer memory or a magnetic tape is required for storage, and the computer access time can be excessive, particularly if magnetic tape or other off-line storage is required.

Since the authors are interested in the use of mini- and microcomputers for calculations in the laboratory or more remote locations, programs have been initiated for developing more efficient methods for handling and storing radiation cross section data. The approach of primary interest is the use of empirical data fitting techniques and the method of recent primary interest is that of cubic splines. The authors have worked on determining the molecular stopping cross sections for alpha particles [30] and are presently compiling cubic spline coefficients for all the available literature data [31]. Some work has been completed on the piece-wise fitting of neutron cross sections [32] and further work is in progress in this area using cubic spline fitting [33]. Work is also progressing on using a specialized discontinuous form of cubic splines for fitting X-ray cross section data across the characteristic absorption discontinuities [34]. The cubic spline approach to the storage and handling of cross section data should prove to be very advantageous for the Monte Carlo models to be used in the smaller computers.

5. Specific Monte Carlo Models for Radiation Analyzers

The generic problems specific to the development of Monte Carlo models for radiation analyzers include those specific to radiation gauges and additional ones brought about by the need to produce spectral responses rather than simple counting rates. In particular the need for spectral responses accentuates the problem of obtaining accurate model predictions for a reasonable number of histories. Significant variance reduction methods are essential in this case.

The authors have worked on the use of specific Monte Carlo models for two radiation analyzers; energy-dispersive X-ray fluorescence analyzers and neutron capture prompt gamma-ray analyzers. The first of these is an ideal problem for solution by Monte Carlo in many respects. For a wide range of the characteristic X-ray energies for many elements, the dominant interactions in matter are the photoelectric absorptions that give rise to the characteristic X rays. This means that essentially every interaction gives rise to the phenomenon of interest. Moreover, the X rays are produced isotropically so that complicated angular distributions need not be treated. The Monte Carlo approach easily accounts for the primary, secondary, and tertiary fluorescence produced in samples composed of multiple elements.

A series of papers describing the Monte Carlo models that have been developed for X-ray fluorescence analyzers have been published [35—48]. These models include X-ray fluorescence analyzers excited by: (1) the monoenergetic photons from annular radiosiotope sources, (2) the continuous spectrum from a conventional X-ray tube, and (3) the spectrum from a secondary fluorescer excited by an X-ray tube. Recent models include the source backscatter portion of the spectrum and contributions to the characteristic X-ray intensities by photons that have first undergone coherent or incoherent scattering. The most recent papers [47, 48] describe Monte Carlo models that attempt to predict the entire X-ray fluorescence spectrum and use it in the analysis. A typical predicted spectrum using this model is compared to experimental results in Fig. 1. This has been done for both homogeneous and heterogeneous ternary samples. A very general Monte Carlo program is presently under development that can be used for calculations involving samples that contain as many as 20 elements [49].

More recently work has been initiated on a specific Monte Carlo model for neutron capture prompt gamma-ray analyzers [32]. The initial model was for a bore-hole logging sonde configuration using a ^{252}Cf source and a cylindrical three-inch by three-inch NaI(Tl) detector. The experimental results obtained for verification of this model were for a range of aqueous solutions of sodium chloride. The Monte Carlo predicted results are compared to experimental results in Table 1. The major variance reductionmethod used in this model consisted of simply following the emitted neutron until it was absorbed and then forcing all the characteristic gamma rays of interest to be emitted and to traverse the sample without further interaction in reaching the detector and being detected. This approach eliminated

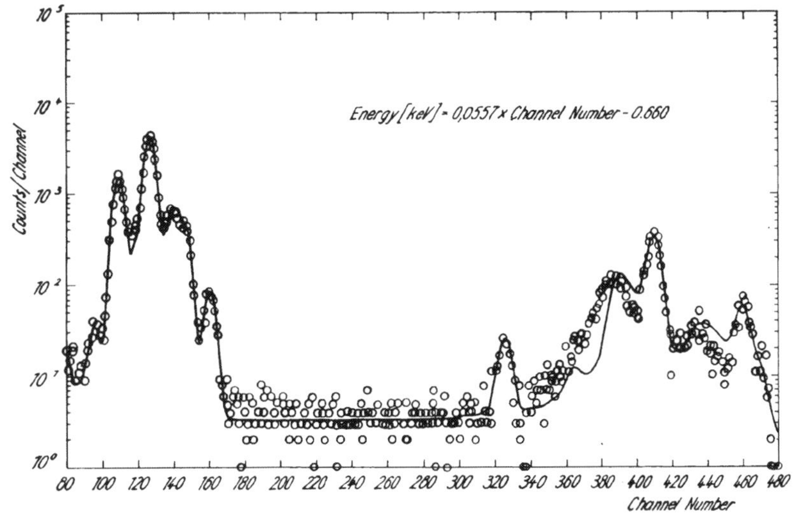

Fig. 1. A comparison of Monte Carlo predicted and experimental pulse-height XRF Spectra for a stainless steel sample excited by a ^{109}Cd source

Tab. 1. Comparison of relative responses predicted by the Monte Carlo model with experimental results for NaCl solutions for a neutron capture prompt gammaray analyzer

Sodium Chloride Concentration %	Relative Counting Rate for Chlorine 6.111 MeV Photopeak and Single Escape Peak from the 6·62 MeV Gamma Ray	
	Experimental	Monte Carlo Model
0·801	0·491 (0·011)*)	0·4265 (0·0050)
1·196	0·716 (0·012)	0·6261 (0·0067)
1·589	0·889 (0·012)	0·8174 (0·0083)
1·978	1·000 (0·013)	1·0000 (0·0101)
2·365	1·201 (0·014)	1·1760 (0·0120)
2·748	1·381 (0·016)	1·3449 (0·0142)
3·128	1·453 (0·015)	1·5063 (0·0168)
	Relative Counting Rate for Hydrogen 2·223 MeV Photopeak	
0·402	1·3034 (0·001)	1·3506 (0·0104)
0·801	1·2188 (0·001)	1·2409 (0·0065)
1·196	1·1414 (0·001)	1·1481 (0·0038)
1·589	1·0701 (0·001)	1·0690 (0·0018)
1·978	1·0000 (0·001)	1·0000 (0·0005)
2·365	0·9572 (0·001)	0·9409 (0·0014)
2·748	0·8992 (0·001)	0·8884 (0·0023)
3·128	0·8556 (0·001)	0·8418 (0·0030)

*) The values in parantheses are the standard deviations due to counting rate fluctuations.

the need for first constructing a detailed neutron flux map within the sample. This work is being extended to the development of a more sophisticated Monte Carlo model which will include the use of the last-flight estimator technique [50] and the development of another Monte Carlo model for a conveyor belt geometry [51].

For both of these applications the specific Monte Carlo models that have been developed yield acceptable accuracies in very reasonable computer run times of less than five minutes on the IBM 370/165 computer. Work is progressing on reducing the computer memory requirements for the cross section data so that these models could be used on small mini — or even microcomputers.

6. Discussion and Conclusions

Monte Carlo models are capable of providing accurate solutions to problems involving radiation gauges and analyzers for reasonable costs (computer run times). It is the opinion of the authors that the Monte Carlo method should be adopted more widely, especially by those involved in the design and use of radiation gauges and analyzers. Specific Monte Carlo models (as opposed to existing general purpose codes) have the additional advantages of: (1) providing considerable insight into the problem at hand to the model developer, (2) being able to replace tedious, costly, and time consuming experiments, and (3) in some cases being able to provide responses for hypothetical samples that cannot be otherwise obtained directly by experiment.

One area of application that appears to be quite promising is in the use of Monte Carlo models on conjuction with detector response functions to provide "synthetic" responses to specified standard samples which can provide a non-experimental method of calibration. This approach would be particularly useful in those cases where standard samples of the desired composition and density were either not available or could not be easily maintained without deterioration.

Acknowledgements

The authors gratefully acknowledge the support of the National Science Foundation for the travel grant (NSF Grant CPE-8211899) allowing this paper to be presented by one of the authors and to the three members (*Exxon, DuPont*, and *Kennedy Van Saun*) of the Center for Engineering Applications of Radioisotopes Associates Program for their general support of our research.

Received February 1, 1983

References

[1] *E. D. Cashwell, C. J. Everett*, A Practical Manual on the Monte Carlo Method for Random Walk Problems, Pergamon Press, Inc., New York 1959

[2] *L. L. Carter, E. D. Cashwell*, "Particle-Transport Simulation with the Monte Carlo Method, Technical Information Center, Office of Public Affairs, U.S. Energy Research and Development Administration, TID-26607, 1975

[3] *Y. A. Shreider* (Ed.), The Monte Carlo Method, Pergamon Press, Inc., New York 1966

[4] *J. M. Hammersley, D. C. Handscomb*, Monte Carlo Methods, John Wiley & Sons, Inc., New York 1964

[5] *J. Spanier, E. M. Gelbard*, Monte Carlo Principles and Neutron Transport Problems, Addison-Wesley Publishing Company, Inc., Reading, Massachusetts, 1969

[6] *N. M. Schaeffer* (Ed.), Reactor Shielding for Nuclear Engineers, U. S. Atomic Energy Commission, TID-25951, 1973

[7] *E. J. McGrath, S. L. Basin, R. W. Burton, D. C. Irving, S. C. Jaquette, W. R. Keller, C. A. Smith*, "Techniques for Efficient Monte Carlo Simulation", Volumes I, II and III, ORNL-RSIC-38, Oak Ridge National Laboratory 1975

[8] O5R Code User's Manual, Radiation Shielding Information Center, Oak Ridge National Laboratory 1967

[9] *D. C. Irving, R. M. Freestone*, Jr., *F. B. Kam*, "A General Purpose Monte Carlo Neutron Transport Code", ORNL-3622, Oak Ridge National Laboratory 1965

[10] *W. J. Lippold*, "Monte Carlo Simulation of Neutron Thermalization in Soils", unpublished M. S. thesis, Department of Nuclear Engineering, North Carolina State University 1969

[11] *W. J. Lippold, A. Carnesale, R. P. Gardner*, Nuclear Engineering and Design **10** (1969) 373

[12] *D. K. Trubey, S. K. Penny, M. B. Emmett*, "OGRE, A Monte Carlo System for Gamma-Ray Transport Studies, Including an Example (OGRE-P1) for Transmission through Laminated Slabs", ORNL-3805, Oak Ridge National Laboratory 1966

[13] *D. K. Trubey, M. B. Emmett*, "OGRE-G, An OGRE System Monte Carlo Code for the Calculation of Gamma-Ray Dose at Arbitrary Points in an Arbitrary Geometry", ORNL-TM-1212, Oak Ridge National Laboratory 1966

[14] *W. C. Hopkins*, "Monte Carlo Simulation of the Spectral Response of a Gamma-Ray Desnsity Gauge", unpublished M. S. thesis, Department of Nuclear Engineering, North Carolina State University 1972

[15] *W. C. Hopkins, R. P. Gardner*, Nuclear Engineering and Design **24** (1973) 332

[16] *R. P. Gardner, W. L. Dunn, F. H. McDougall, W. J. Lippold*, "Optimization of Density and Moisture Content Measurements by Nuclear Methods", National Cooperative Highway Research Program Report 125, Highway Research Board, National Research Council, National Academy of Sciences — National Academy of Engineering, from North Carolina State University 1971

[17] *E. A. Straker, W. H. Scott*, Jr., *N. R. Byrn*, "MORSE General Purpose Monte Carlo Multigroup Neutron and Gamma-Ray Transport Code with Combinatorial Geometry", RSIC Computer Code Collection Report CCC-127 D, Oak Ridge National Laboratory 1972

[18] *R. A. Straker, G. L. Simmons*, "An Engineers Guide to the Use of Monte Carlo in Shielding Studies", SAI Services, Inc., Huntsville, Alabama, January 15, 1975

[19] *J. Gyurcsak*, International Atomic Energy Agency Post-Doctoral Fellow, Center for Engineering Applications of Radioisotopes, North Carolina State University 1980—1981

[20] *R. P. Gardner, A. Carnesale*, Nuclear Instruments and Methods **73** (1969) 228

[21] *R. P. Gardner, K. Verghese*, Nuclear Instruments and Methods **93** (1971) 163
[22] *K. Verghese, R. P. Gardner, R. M. Felder*, Nuclear Instruments and Methods **101** (1972) 391
[23] *W. L. Dunn, R. P. Gardner*, Nuclear Instruments and Methods **103** (1972) 373
[24] *G. R. Beam, L. Wielopolski, R. P. Gardner, K. Verghese*, Nuclear Instruments and Methods **154** (1978) 501
[25] *R. P. Gardner*, Nuclear Instruments and Methods **138** (1976) 287
[26] *A. Porelli, R. P. Gardner*, Nuclear Instruments and Methods **159** (1979) 177
[27] *L. Wielopolski, R. P. Gardner*, Advances in X-Ray Analysis **22** (1979) 317
[28] *L. Wielopolski, R. P. Gardner*, Nuclear Instruments and Methods **165** (1979) 297
[29] *R. P. Gardner, J. M. Doster*, Nuclear Instruments and Methods **198** (1982) 381
[30] *F. Wenger, R. P. Gardner, K. Verghese*, Health Physics **25** (1973) 67
[31] *M. J. Jensen, R. P. Gardner*, to be submitted to Health Physics
[32] *T. C. Clark, R. P. Gardner, K. Verghese*, Nuclear Instruments and Methods **193** (1982) 365
[33] *A. Almasoumi, R. P. Gardner, K. Verghese*, to be submitted to Nuclear Instruments and Methods
[34] *A. Yacout, R. P. Gardner, K. Verghese*, to be submitted to Nuclear Instruments and Methods
[35] *R. P. Gardner, A. R. Hawthorne*, X-Ray Spectrometry **4** (1975) 138
[36] *A. R. Hawthorne, R. P. Gardner*, Transactions of the American Nuclear Society, Supplement No. 3, **21** (1975) 38
[37] *A. R. Hawthorne, R. P. Gardner*, Analytical Chemistry **47** (1975) 2220
[38] *A. R. Hawthorne, R. P. Gardner, T. G. Dzubay*, Advances in X-Ray Analysis **19** (1975) 323
[39] *A. R. Hawthorne, R. P. Gardner*, Analytical Chemistry **48** (1976) 2130
[40] *A. R. Hawthorne, R. P. Gardner*, "Chapter 18. Monte Carlo Applications to the X-Ray Fluorescence Analysis of Aerosol Samples", in X-Ray Fluorescence Analysis of Environmental Samples", T. G. Dzubay (Ed.), Ann Arbor Science Publishers, Inc., Ann Arbor, Michigan 1977, p. 209
[41] *R. P. Gardner, L. Wielopolski, K. Verghese*, Atomic Energy Review, **15** (1977) 701
[42] *R. P. Gardner, L. Wielopolski, J. M. Doster*, Advances in X-Ray Analysis **21** (1978) 129
[43] *R. P. Gardner, L. Wielopolski, K. Verghese*, Journal of Radioanalytical Chemistry **43** (1978) 611
[44] *R. P. Gardner, J. M. Doster*, Advances in X-Ray Analysis **22** (2979) 343
[45] *F. Arinc, R. P. Gardner*, Journal of Radioanalytical Chemistry **54** (1979) 221
[46] *J. M. Doster, R. P. Gardner*, Advances in X-Ray Analysis **23** (1980) 117
[47] *J. M. Doster, R. P. Gardner*, X-Ray Spectrometry **11** (1982) 173
[48] *J. M. Doster, R. P. Gardner*, X-Ray Spectrometry **11** (1982) 181
[49] *A. Yacout*, unpublished Ph. D. thesis in preparation
[50] *A. Almasoumi*, unpublished Ph. D. thesis in preparation
[51] *Y. L. Yuan*, unpublished Ph. D. thesis in preparation

Isotopenpraxis **19** (1983) 10, S. 335–338

Evaluation of Autoradiograms Using a Microcomputer*)

W. Birkholz, M. Steinert

(Central Institute of Isotope and Radiation Research, Academy of Sciences of the GDR)[1])

An equipment AURAS for evaluation of autoradiograms was developed. It consists of a digitalphotometer and the densitron system (television scanner) for digitalisation of pictures and a microprocessor for picture processing and storing of data.
The digitalphotometer permits a precise but timeconsuming scanning. The densitron system works quickly with little density classes and pseudocolouring of pictures.
For the evaluation of autoradiograms with the microcomputer a programsystem MARAUS exists. It works in a dialogeous regime and has following functions:
data transfer between the scanners and the computer; estimation of density histograms and areas of different density classes in the picture; calculation of average of densities; printing of density profiles or classifite pictures; transfer of data to external medias such as punched tape or magnetic cassette for piture processing on bigger computers. The possibilities of using the equipment for evaluation of autoradiograms are demonstrated.

Keywords

autoradiography; image processing; microprocessor; picture evaluation

Introduction

Autoradiographic methods for the imaging of substance distributions are used for solving problems in the fields of metallurgy, semiconductor industry, chemical industry, biology, medicine, and others. Besides qualitative imaging quantitative determinations of substance distributions are becoming more and more interesting. It is possible to get quantitative results from autoradiograms by the calculation of average densities, density sums, density histograms or the analysis of object structures. This requires more sophisticated measuring devices. Therefore a number of picture processing instruments were deloped. They work on two different principles:

- measuring of the density point by point, digitization, computer processing
- classification of the density into a few classes (equidensitometry) and calculation of the area of each density class.

At the Central Institut of Isotope and Radiation Research, Leipzig a microprocessor-supported system for the evaluation of autoradiograms AURAS was deloped based on commercial devices such as Zeiss rapid photometer and the densitron system.

*) Presented at the 2nd Working Meeting "Radioisotope Application and Radiation Processing in Industry", 28th September to 1st October 1982, Leipzig

[1]) Address: DDR-7050 Leipzig, Permoserstr. 15

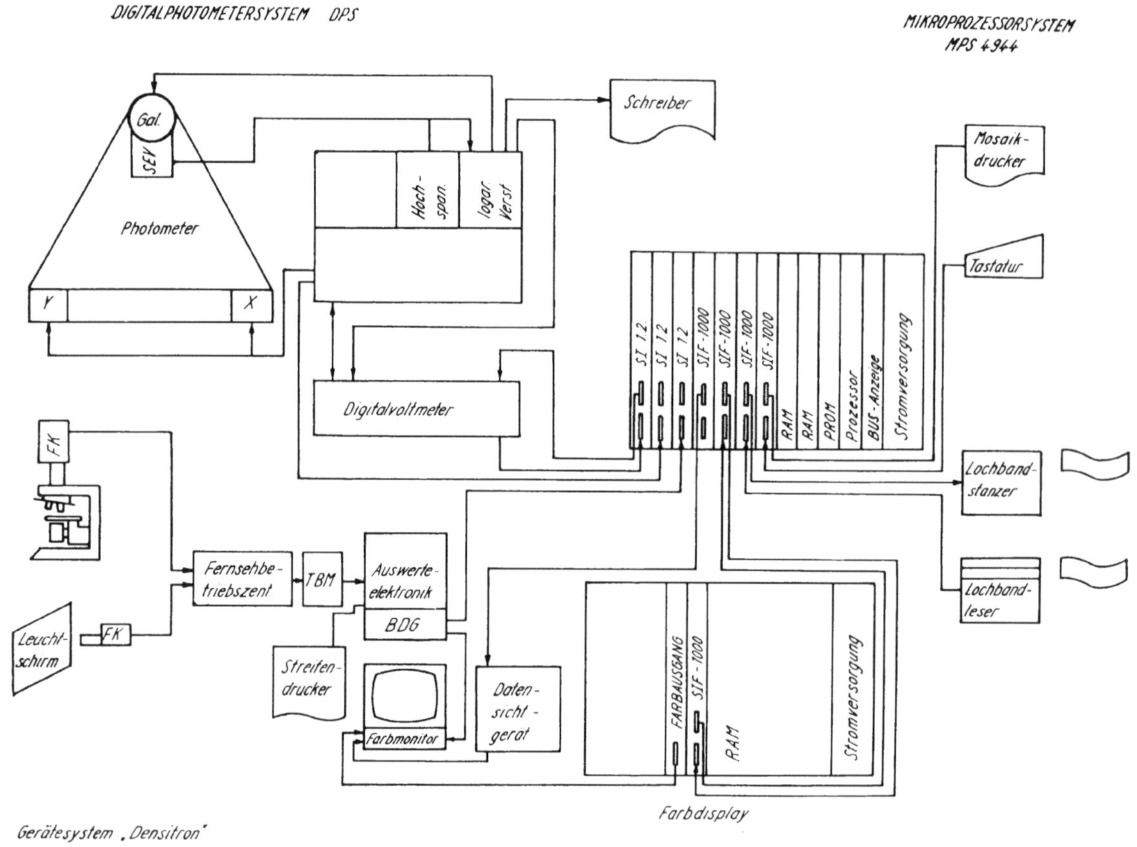

Fig. 1. Block diagram of the AURAS system

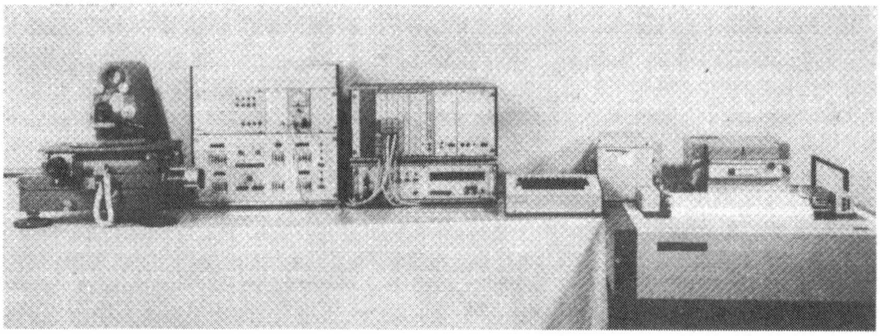

Fig. 2. Digital photometer system DPS with microcomputer MPS

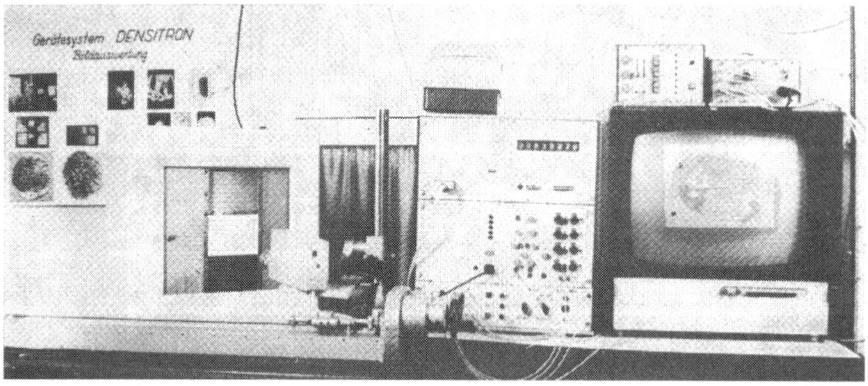

Fig. 3. Densitron system with BDG scanner

Instrumentation

The central unit of the autoradiogram evaluation system is a microprocessor system MPS of the Central Institut of Nuclear Research, Rossendorf. It is used with a processor card, input-output cards, 21 kbyte RAM and 7 kbyte EPROM cards. The input of picture data is accomplished either by the modified Zeiss rapid photometer or the densitron system with the BDG scanner, developed at the ZfI and manufactured by the Centre of the manufacture of scientific instruments, Berlin.

For communication between the operator and the system AURAS an alphanumerical display is used.

The Zeiss rapid photometer is modified to become a digital photometer by the use of stepping motors with electronic devices. The stage with the autoradiogram can be moved automatically in the x- and y-directions with freely chosen numbers of measuring points and various distances between these points. For the density measurements a photomultiplier with a logarithmic amplifier and ADC for the computer input is used.

The densitron system was supplemented by a BDG scanner to couple it with the microcomputer. The picture is scanned by a television camera. The signal of the camera is divided into 6 classes depending on the density distribution in the autoradiogram. It is represented in pseudocolours on a colour display. By the aid of the BDG scanner a scanning grid with freely chosen position, size and fineness is superimposed on the television picture and the information about which colour a particular point has is transferred to the microcomputer. A digital picture is formed with 7 classes (black equals no colour) and a maximum of 400 × 300 points. Detailed descriptions of the device are found in the literature.

Fig. 1 shows the block diagram and Fig. 2 and 3 photographs of the autoradiogram evaluation system ARAUS.

Program system

For microprocessor-supported autoradiogram evaluation the program system MARAUS was created. It allows dialog oriented work with the AURAS system. Communication between the operator and the microcomputer system MPS is accomplished by keyboard and a display. A hard copy device can be used optionally.

The rough program structere of MARAUS is shown in Fig. 4. The program system has a modular structure. It has a maximum of 16 basic commands. At present only 5 basic commands are realised:

A – assignment of the starting point for picture storage in the RAM.

B – picture output on hard copy device with a maximum of 16 gray symbols.

C – calculation of colour areas or histograms.

D – storage of picture data on peripherial storages.

L – loading of picture data from the densitron system, digital photometer or peripherial storage.

Other functions can easily be added without changing the software already present.

The peripherial devices of the microcomputer are supported by monitor software. The program system MARAUS was joined to the monitor as an extension. This gives the advantage of using the utility programs of the monitor for testing user programs.

The program system was written in the assembler language MAPS K1520 and needs a storage of 7 kbyte (EPROM)

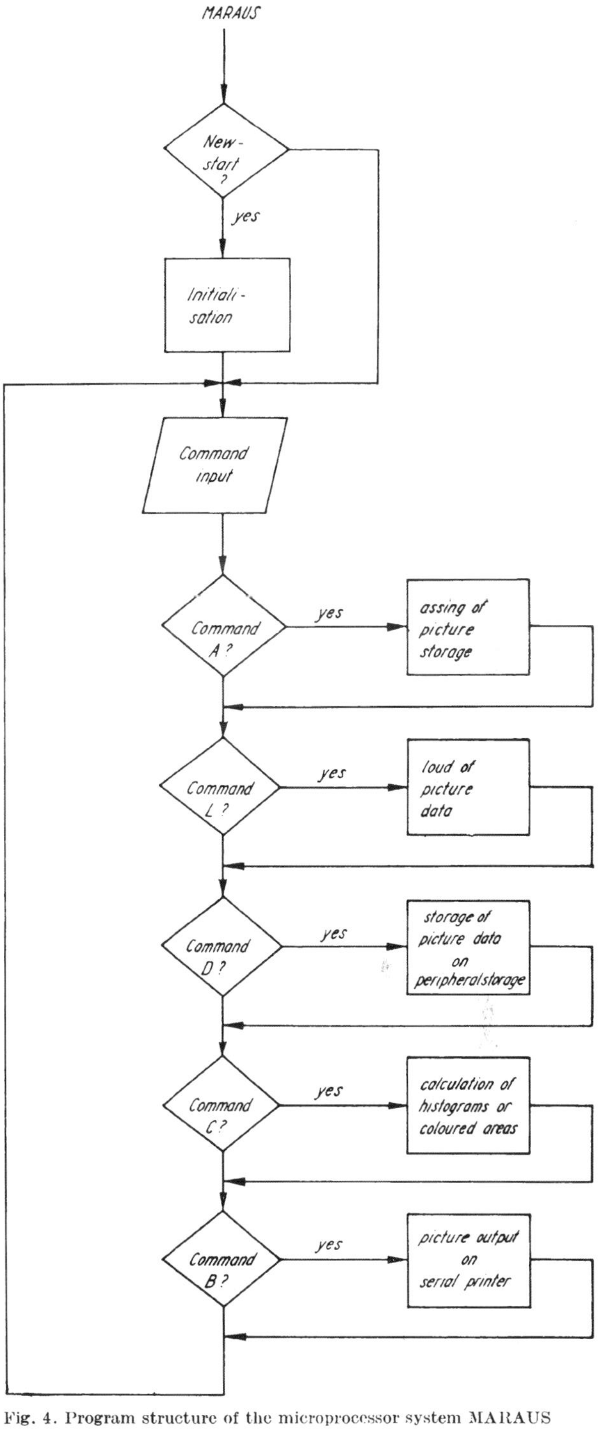

Fig. 4. Program structure of the microprocessor system MARAUS

inclusive of the monitor. Together with the stack a scratchpad of 1 kbyte (RAM) is necessary. In addition there is a picture storage of 20 kbyte RAM at present.

Particular care was taken in making the programs for the keyboard input. They must ensure uniquess and allow the operator to make corrections and remarks.

Example of application

Fig. 5 shows the copy of an autoradiogram of the distribution of a ^{32}P labelled herbicide in a pea plant. The autoradiogram was scanned with the densitron system and processed with the program system MARAUS. Fig. 6 presents the printed record of the working of MARAUS.

The commands used were:

L: loading of data from the densitron system;
C: calculation of colour areas;
B: printing of the picture with the serial printer.

The command L may be specified by the declaration fo the input medium:

D: densitron system;
P: digital photometer;
M: punched tape.

Fig. 5. Autoradiogram of the distribution of a ^{32}P labelled herbicide in a pea plant

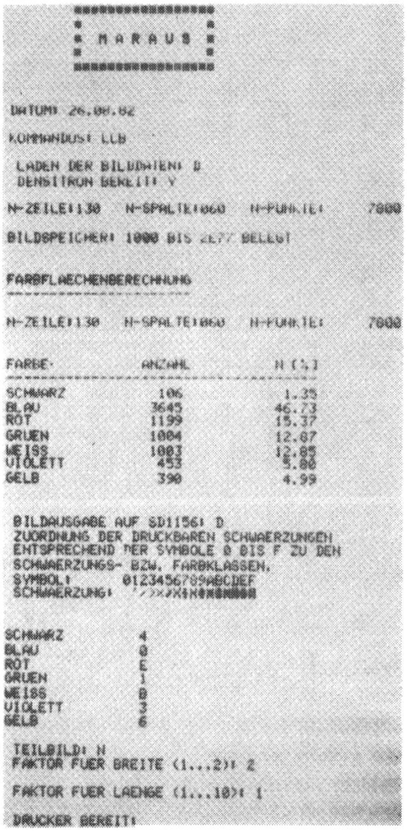

```
* M A R A U S *

DATUM: 26.08.82

KOMMANDOS: LLB

 LADEN DER BILDDATEN: D
 DENSITRON BEREIT: Y

N-ZEILE:130   N-SPALTE:060   N-PUNKTE:      7800

BILDSPEICHER: 1000 BIS 2E77 BELEGT

FARBFLAECHENBERECHNUNG
----------------------

N-ZEILE:130   N-SPALTE:060   N-PUNKTE:      7800

FARBE          ANZAHL        H (%)
----------------------------------
SCHWARZ          106          1.35
BLAU            3645         46.73
ROT             1199         15.37
GRUEN           1004         12.87
WEISS           1003         12.85
VIOLETT          453          5.80
GELB             390          4.99

 BILDAUSGABE AUF SD1156: D
 ZUORDNUNG DER DRUCKBAREN SCHWAERZUNGEN
 ENTSPRECHEND DER SYMBOLE 0 BIS F ZU DEN
 SCHWAERZUNGS- BZW. FARBKLASSEN.
 SYMBOL:       0123456789ABCDEF

SCHWARZ        4
BLAU           0
ROT            E
GRUEN          1
WEISS          8
VIOLETT        3
GELB           6

 TEILBILD: N
 FAKTOR FUER BREITE (1...2): 2

 FAKTOR FUER LAENGE (1...10): 1

 DRUCKER BEREIT:
```

Fig. 6. Printed record of the working of the MARAUS system

When the densitron system is used at first the number of rows and colums chosen on the BDG scanner and then the picture data are transferred.

Command C serves to calculate the absolute and relative frequencies of the colours in the picture and to print them in the form of a table. If the scanning of the autoradiogram is done with the digital photometer the C-command allows the calculation of a density histogramm with a chosen number of classes.

For printing computer-processed pictures with the serial printer the following possibilities exist:

- picking out parts of the picture by declaration of the start and the end address;
- declaration of expansion coefficient in the row and colum directions; In this way it is possible to correct the geometrical distortion due to the printer sign.
- assignment of density symbols to the density classes of the picture. 16 gray symbols can be repsented.

A part of the autoradiogram is represented by means of the serial printer in Fig. 7.

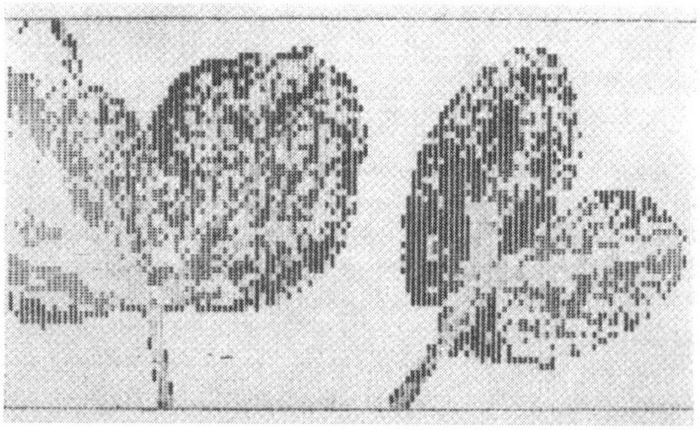

Fig. 7. Representation of the autoradiogram (fig. 5) by means of a serial printer

Conclusion

The system AURAS is a device for the digital evaluation of autoradiograms and exact density measurements with the digital photometer as well as a coupling of the densitron system to a computer. The device for data output of the densiton system, the BDG scanner, is manufactured in series by the manufactures of the densitron system.

The program system MARAUS is dialog oriented and permits simple picture processing. It can easily be extended by special programs of the user.

Received January 1, 1983
Accepted in revised form February 23, 1983

References

[1] *H. Kazmierczak*, Erfassung und maschinelle Verarbeitung von Bilddaten Berlin: Akademieverlag 1980
[2] *H. Baumbach, K. Freyer, G. Kummer, W. Birkholz*, Isotopenpraxis **12** (1976) 81; 145
[3] *K. Freyer, H.-Ch. Treutler, W. Birkholz, D. Slams, H. Friedrich, H. Koepernick*, radio · fernsehen · elektronik **29** (1980) 551
[4] *W. Birkholz, W. Uhlmann, O. Schlimpert, K. Freyer*, radio · fernsehen · elektronik **29** (1980) 644
[5] *M. Kahlenbach, K.-H. König, K.-W. Leege, D. Pröhl*, Messen · Steuern · Regeln **21** (1978) 488
[6] *W. Birkholz, K. Freyer*, Radioanal. Chem., **79** (1983) 245
[7] *W. Birkholz, M. Steinert, K. Freyer*, MARAUS — ein Programmsystem zur mikrorechnergestützten Autoradiogrammauswertung. Tagungsberichte der AdL, **1983**, 169

Isotopenpraxis 19 (1983) 10, S. 339 – 341

Radiotracer Investigations of Benefication Copper Ore in the Industrial Flotation Process*)

L. Petryka, K. Przewlocki

(Institute of Physics and Nuclear Techniques Academy of Mining an Metallurgy, Cracow)[1])

Optimization of Copper ore benefication requires identification of the whole process, in which flotation is the main operation. Natural polymetallic ore in which activity of ^{64}Cu was induced served like a tracer. Subject of detailed measurements was a 14 chamber commercial flotation machine. As a result mean residence time for different granulometric fractions of natural copper sulphide was determined for all parts of flotation machine, as well as the efficiency of flotation.

Keywords

copper 64; copper ores; flotation; ore enrichment; tracer techniques

1. Introduction

Over 20 mln. tons of copper ore are processed in Poland each year. Average copper content in it varies about 2 per cent and must be increased 10 times before sending it to the smelting works. After hoisting copper ore from the mine it must be suitably ground, selected and finally by means of flotation, copper content must be increased to the certain limit in the appropriate dressing plant. Fig. 1 shows the simplified diagram of the process.

In the largest Polish polymetallic deposit (LGOM-Lower Silesia) copper ores posses variable petrographic composition and appear like intrusions of copper minerals such as chalcosine (CuS) and chalcopirits ($CuFeS_2$) in shale,

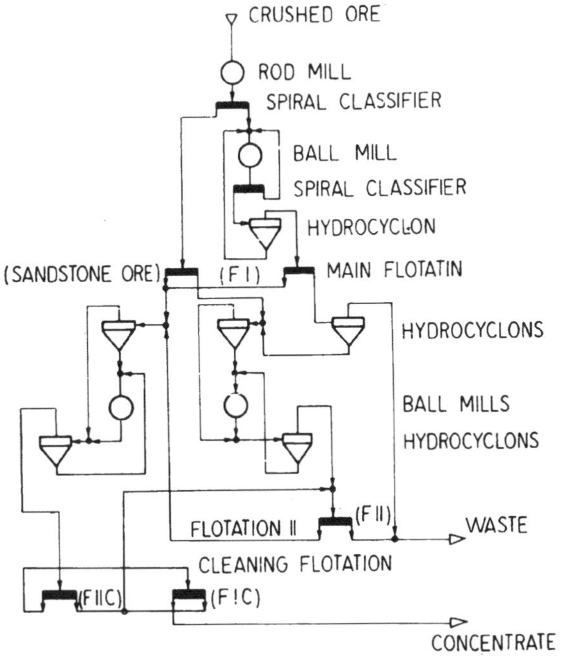

Fig. 1.

*) Presented at the 2nd Working Meeting "Radioisotope Application and Radiation Processing in Industry", 28st September to 1th October 1982, Leipzig

[1]) Address: 30-059 Cracow, al. Mickiewicza 30, Poland

limestones and sandstones. As a consequence sharp requirements must be imposed to the grinding and classification processes.

Similarly control of benefication in the flotation machines must be performed in the way allowing, in spite of the variable composition of the feed, keeping constant amount of copper at the outlet and limitation losses in the tailings to the indispensable minimum.

The last stages of dressing: drying of concentrate and removing tailings to the ponds are the most arduous for the environment.

Identification of the copper ore dressing processes by means of radioactive tracers has a 10 year old tradition [1]. Process of grinding is rather well recognized. It was established that in the rod mill mean residence time is longer for minerals less susceptible for grinding and rises with the size of the grains feeding mill [2].

Results presented in a paper deal with the flotation process, main operation of the whole dressing process. The main subject of interest was intensity of transfer of the copper minerals to the concentrate in the subsequent part of the flotation machine.

2. Method of measurements

The measurements were based on the copper isotope ^{64}Cu used as a tracer, activity of which was induced in the natural chalcosine. In order to establish flotativity of the copper sulphides as a function of a grain size 4 granulometric fractions of ore were prepared: <8 µm, 8–20 µm, 20–40 µm and >40 µm. 18 ore samples, 0·1 kg each was activated during 1·5 min in the neutron flux $5 \cdot 10^{16}$ neutrons/sq · ms and cooled 50 h to avoid contribution of ^{56}Mn and other short living radioisotopes. Finally its activity was kept at the level 18 MBq. To decrease natural background of the scintillation probes only a piece of the whole gamma spectrum at the range of 0·51 MeV (β^+ anihilation peak) emited by ^{64}Cu isotope [3], was recorded.

All together 6 probes were used, working in the 6 separate measuring lanes. Breaks through curves were analysed and mean residence time was determined at their base, for each operation of the process. At the same time after proper calibration of the probes it became possible to determine copper separation coefficients between concentrate and tailings in the examined parts of flotation machine.

Some technological parameters, determining establishing coordinates of state of the investigated process were measured as well.

3. Tracer mass balance

Measurements were performed at the selected flotation machine belonging to the main flotation stream of the Polkovice Dressing Plant. Fig. 2 show manner of placing the measuring probes. Probes 3, 5 and 6 were placed below the gutters conducting copper concentrate gathered in the separate parts of the flotation aggregate. Probes 2 and 4 were laid over the pipe-lines carrying subsequently tailings and concentrate. Thus probe No 5 detected tracer gathered in the 3-d and 4-th part of the machine, below probe No 4 tracer gathered on the whole length, below probe No 3 tracer from the 1-st part and below probe No 6 tracer gathered in the 4-th part and finally probe No 2 amount of the tracer coming to the tailings. 6 probes used at the same time made possible not only determination flow rate of tracer at the outlets of concentrate and tailings, but also intensity of its transit to the concentrate in the particular parts of the machine.

After proper calibration it was relatively easy to measure concentration of the tracer at the inlet and two outlets of the flotation machine. It became evident that recognition of processes taking part inside is more complex. This was mainly due to the construction of the machine itself; probes placed over the gutter conducting concentrate from the separate cells, recorded at the same time, radiation coming from the gutter and from the fluid inside of the main vessel (probes 3, 5, 6). Mass balance procedure made it possible to solve the problem.

Graph at the Fig. 3 illustrates it in another way; coefficients $K_1 - K_4$ have to be determined.

Calibration of the probe performed in the laboratory gave the results in Tab. 1, where $m = p_i/p_2$, p_i, p_2 integrated counts for the investigated and reference probe. Dimensionless coefficients u express relative efficiency taking into account different geometries and velocities of the flowing medium, all compared with the working condition of the probe No. 2

To derive mass balance, special mass unit was introduced — "1 r.m.u." (relative mass unit) corresponding to such mass of the tracer which flowing with the speed of tailings (measured by reference probe) cause 1 count in the pertinent other probe.

Mass balance was calculated using following formulas (mass expressed in the r.m.u.):

Mass of the tracer coming to the tailings:

$$M_t = \sum_i N_{i2} u_2 / m_2$$

Tab. 1. Calibration of the probes

Probe No	m	u
2	1	1
3	·809	·91
4	·691	2·55
5	·680	1·73
6	1·089	1·55

Tab. 2.

Tracer		Separation of Copper						Efficiency per one cell			
Fraction µm	Kind of Mat.	Conc. %	Tailings %	Conc.₁ %	Conc.₂ %	Conc.₃ %	Conc.₄ %	Conc.₁ —	Conc.₂ —	Conc.₃ —	Conc.₄ —
<8	Chalcosine	93 ± 1	7 ± 1	11 ± 1	53 ± 4	22 ± 3	7 ± 1	0.5 ± 0.1	2.7 ± 0.2	1.2 ± 0.2	1.0 ± 0.1
	concentrate	99 ± 1	1 ± 1	16 ± 1	67 ± 4	12 ± 2	4 ± 1	1.1 ± 0.1	5.8 ± 0.2	1.2 ± 0.2	1.0 ± 0.1
8–20	Chalcosine	89 ± 1	11 ± 1	12 ± 1	48 ± 4	24 ± 3	6 ± 1	0.6 ± 0.1	2.9 ± 0.2	1.6 ± 0.2	1.0 ± 0.1
	Concentrate	98 ± 1	2 ± 1	15 ± 3,5	75 ± 4	4 ± 0.5	2 ± 0.5	2.2 ± 0.8	14 ± 5	1.0 ± 0.2	1.0 ± 0.1
20–40	Chalcosine	96 ± 1	4 ± 1	8 ± 1	65.5 ± 3.5	17.5 ± 3.5	5 ± 1	0.45 ± 0.05	4.8 ± 0.2	1.4 ± 0.3	1.0 ± 0.1
>40	Chalcosine	94 ± 1	6 ± 1	10 ± 2	45 ± 4	32 ± 1	7 ± 1	0.4 ± 0.1	2.5 ± 0.5	1.8 ± 0.2	1.0 ± 0.1
	Concentrate	95 ± 1	5 ± 1	10 ± 1	55 ± 3	25 ± 2	5 ± 1	0.5 ± 0.1	3.5 ± 0.2	1.7 ± 0.2	1.0 ± 0.1

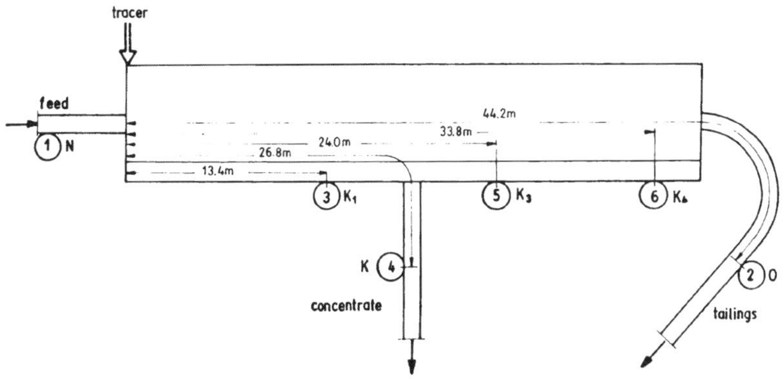

Fig. 2. Measurement scheme of the flotation machine

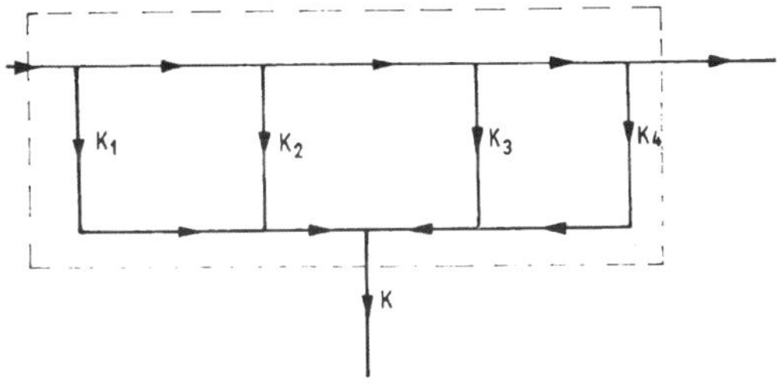

Fig. 3. Flow-graph for the tracer in the flotation process

Mass of the tracer coming to the concentrate:

$$M_c = \sum_i N_{i4}u_4/m_4$$

Mass of the tracer coming to the concentrate in the 4-th part of the flotation machine:

$$M_{c4} = \sum_i N_{i6}u_6m_6 - wM_t$$

Mass of the tracer coming to the concentrate in the 3-rd part:

$$M_{c3} = \sum_i N_{i5}u_5/m_5 - wM_t - (1 + w)\, M_{c4}$$

Mass of the tracer coming to the concentrate in the 1-st part

$$M_{c1} = \frac{\sum_i N_{i3}u_3m_3 - wM_t - M_c}{1 - w}$$

Mass of the tracer coming to the concentrate in the 2-nd part:

$$M_{c2} = M_c - M_{c1} - M_{c3} - M_{c4}$$

where $w = {\cdot}015$ coefficient was established experimentally. It was presented that probes 3, 5 and 6 placed at the gutter conducting concentrate, record 1·5 per cent of the tracing flowing inside the flotation machine.

Mass balance principle requires that the mass of the tracer injected "M_{in}" has to be the sum of "M_c" and "M_t"

Assuming "M^{in}" equal 100 per cent of the tracer Tab. 2 presents values of subsequent separation coefficients.

To get more readable picture and taking into account that each part of the flotation machine contains different number of the flotation chambers Tab. 2 depicts additionally efficiency of the particular cells related to the efficiency of the part 4.

4. Conclusions

On the basis of these measurements one can draw following conclusions:

- grinding has to be controlled more rigorously to eliminate too coarse and too fine grains in the feed
- finest grains flowing through the flotation machine behave similarly as water and only a small part of it become concentrate
- it looks sensibly to apply additional classification of grains at the tailings, to select coarse grains and direct them back to the mill.

5. Acknowlegements

Author acknowledges financial support of the Institute of Nuclear Research, kind cooperation of the dressing plant staff and the cooperation of Mr. Mr. *Z. Stęgowski*, *A. Stós*, *L. Furman*, *M. Franaszek* and Miss *M. Porębska* for taking part in the field measurements and data processing.

Received February 15, 1983

References

[1] *K. Korbel* et al. – Application Possibilities of Identification of Particular Objects of a Copper Ore Preparation Plant. Rep. 817/J Institute of Nuclear Physics, Cracow 1972

[2] *K. Korbel* et al. – Radioisotope Investigations of Grainding and Classification in Copper Ore Dressing Plant Rep. for Institute of Nonferrous Metals in Gliwice, Cracow 1973

[3] *K. Korbel* et al. – Identification of Copper Ore Processing by Means of Radioactive Tracers. IAEA-SM-216/31, Vienna 1977

Isotopenpraxis 19 (1983) 10, S. 342–344

Gamma-Radiolyse von Dimethylglyoxim

H. Langguth, H.-U. Försterling, J. Schmidt

(Zentralinstitut für Isotopen- und Strahlenforschung der AdW der DDR Leipzig)[1])

Die Strahlenresistenz von reinem und an Aktivkohle sorbiertem Dimethylglyoxim (DMG), das für die selektive Abtrennung von Palladium aus Abfallösungen der Kernbrennstoffwiederaufarbeitung geeignet ist, wurde untersucht. Die Bestrahlung erfolgte bei 20–40 °C mit einer ^{60}Co-γ-Bestrahlungsanlage mit Dosisleistungen von 0,6 bis 1,7 Gy/s in einem Dosisbereich von 2 kGy bis 5 MGy. Die Radiolyseprodukte wurden mit verschiedenen Methoden analysiert und die G-werte bestimmt. Es wurde gezeigt, daß DMG bis zu Dosen von etwa 110 kGy relativ strahlenresistent ist. Bei höheren Dosen werden die reaktiven NOH-Gruppen des DMG teilweise zerstört, so daß sie nicht mit Palladium reagieren können. In Verfahren zur Abtrennung von Palladium aus radioaktiven Abfallösungen werden diese Dosen jedoch nicht erreicht. Auf eine unzureichende Strahlenresistenz von DMG zurückzuführende Komplikationen können deshalb ausgeschlossen werden.

Keywords

chemical radiation effect; dimethylglyoxime; gamma radiation; radiation doses; radiolysis; radioactive waste reprocessing; fission products; nuclear fuel solutions; palladium

Einleitung

Für die Gewinnung von Spaltprodukten aus Abfallösungen, die bei der Wiederaufbereitung von bestrahlten Kernbrennstoffen anfallen, kommen neben Extaktions-, Ionenaustausch- und Fällungsprozessen auch Verfahren in Betracht, die auf der Sorption an speziell für diese Aufgabe hergestellten, hochselektiven Sorbenzien beruhen. Solche Sorbenzien können entweder durch gezielte Synthese von Austauschern mit selektiven funktionellen Ankergruppen [1–3] oder durch Adsorption von Ragenzien auf unspezifischen Trägermaterilaien erhalten werden [4–12]. Für die selektive Abtrennung von Palladium ist die Verwendung von mit Dimethylglyoxim modifizierten Aktivkohlen möglich [13].

Dimethylglyoxim (Butandioxim-2,3/DMG) ist allgemein als Fällungsreagens für Nickel und Palladium bekannt. In mineralsaurer Lösung ist der Komplexbildner für Palladium selektiv.

Ziel der vorliegenden Arbeit ist es, an Hand von Strahlenresistenzuntersuchungen die Stabilität von DMG zu bestimmen, um seine Anwendbarkeit unter den hochaktiven Bedingungen der Spaltproduktgewinnung aus Kernbrennstoffabfallösungen für die Abtrennung von Palladium abzuschätzen.

Da bisher keine Arbeiten über die Gammaradiolyse von Dimethylglyoxim in der Literatur gefunden wurden, wurde es notwendig, die Strahlenresistenz des DMG unter besonderer Berücksichtigung der reaktiven komplexbildenden NOH-Gruppen experimentell zu untersuchen [14]. Zu diesem Zweck wurde das reine DMG, sowie das auf Aktivkohle sorbierte DMG der γ-Bestrahlung ausgesetzt und anschließend mit verschiedenen Methoden analysiert.

Experimenteller Teil

Die Bestrahlungsexperimente wurden an der Kobalt-60-Gammabestrahlungsanlage vom Typ „RCH-γ-20" mit verschiedenen Dosisleistungen von 0,6 und 1,7 Gy/s im Dosisbereich von 2 kGy bis 5 MGy im Temperaturbereich von 20–40 °C durchgeführt.

Die Bestrahlung der Proben erfolgte in abgeschmolzenen Glasampullen unter Luft und im Vakuum.

Für die analytische Bestimmung der gasförmigen Radiolyseprodukte und des Komplexbildners wurden vier verschiedene Meßmethoden genutzt:

1. Gaschromatographie

Gaschromatographische Bestimmung der Spaltgase an gepackten Säulen mit Wärmeleitfähigkeits- und Flammenionisationsdetektoren am Gaschromatographen des Typs 18.3–4 der Firma "Chromatron".

Nachweis und quantitative Erfassung der Radiolysegase H_2, CH_4, C_2H_6 und C_2H_4 mit den Trennsäulen Molekularsieb 5 A (H_2) und Porapak Q-S (Kohlenwasserstoffe/ durch Gasdosierung über eine Gas-Probenschleife. Berechnung der Konzentrationen an radiolytischen Spaltgasen über Absoluteichungen mit Wasserstoff und Methan und elektronische Auswertung der Peakflächen.

2. Infrarotspektroskopie

IR-spektroskopische Vermessung der komplexbildenden NOH-Gruppe als NO-Valenzschwingung bei 920 cm^{-1} in Abhängigkeit von der Bestrahlungsdosis mit einem IR-Spektrometer des Typs „IR-Specord 75" an 2%igen Lösungen von DMG in Dimethylformamid. Als Vergleichsprobe wurde ein mit Dimethylformamid gefülltes KBr-Prisma verwendet.

3. Protonenkernresonanz

Kernresonanzspektroskopische Vermessung der protonenhaltigen CH_3- und NOH-Gruppen im DMG am 80 MHz Kernresosnanzspektrometer vom Typ BS 487 C (Tesla) an 8%igen Lösungen von DMG in Dimethylformamid – D7 bei Raumtemperatur. Für die Berechnung der Protonensignale im bestrahlten DMG wurden die integrierten Flächen der Resonanzsignale der CH_3- und NOH-Gruppen auf die der CH_3-Gruppen des nicht völlig deuterierten Lösungsmittels DMF – D7 als innerer Standard bezogen.

Die gemessenen Protonensignale enthält Tab. 1.

[1]) Anschrift: 7050 Leipzig, Permoserstr. 15

Tab. 1

Substanz	Struktur-gruppe	chem. Verschiebung ppm
DMG	CH_3	2.01
	NOH	10.48
DMF-D7	CH_3	2.81 und 2.98
	CH	7.90

4. *Fällungsreaktionen*

Die Analyse des strahlenchemischen Abbaus der komplexbildenden Oximgruppe in Abhängigkeit von der Strahlendosis erfolgte auf chemischen Wege durch Fällung mit Pd^{2+} und Ni^{2+}. Die bei einer Dosisleistung von 2 kGy/h an Luft bestrahlten DMG-Proben wurden ausgewogen, in Äthanol gelöst und anschließend wurde die Konzentration an Komplexbildner in der Lösung gravimetrisch bestimmt.

Die Fällungen erfolgten in Anlehnung an die bekannten Vorschriften für die gravimetrische Palladium- bzw. Nickelbestimmung mit DMG [16] entweder mit einer schwach salzsauren Pd(II/chlorid-Lösung mit maximal 3fachem stöchiometrischen Palladium-Überschuß oder mit einer in maximal 7fachem Überschuß zugegebenen Lösung von Nickelammoniumsulfat im Acetatmedium. Die Anwendbarkeit dieser Fällungsverfahren wurde in Vorversuchen geprüft.

Ergebnisse und Diskussion

Die durchgeführten Untersuchungen zeigen, daß DMG durch den Einfluß von Gammastrahlung radiolytisch abgebaut wird. Im Radiolysegas des bestrahlten DMG konnten mittels der Gaschromatographie ab Dosen von 6 kGy Wasserstoff und Methan sowie ab Dosen von 200 bzw. 500 kGy geringe Mengen an Äthylen und Äthan nachgewiesen werden. Die Abspaltung von Wasserstoff nimmt dabei, wie in Abb. 1 gezeigt wird, fast linear mit steigender Dosis zu.

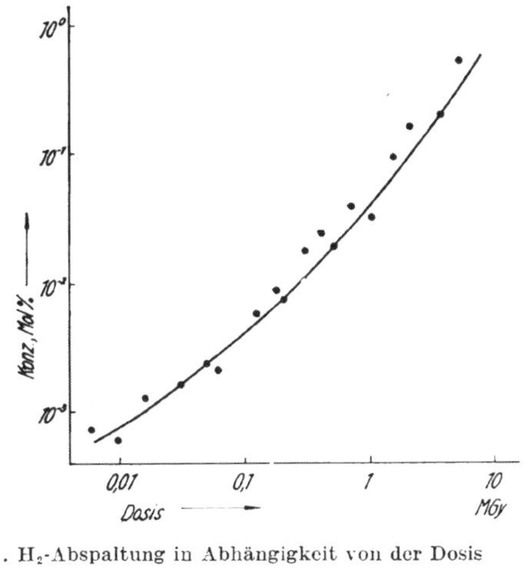

Abb. 1. H_2-Abspaltung in Abhängigkeit von der Dosis

Die NMR-Untersuchungen bestätigen diesen Abbau und zeigen weiterhin, daß offensichtlich die H_2-Abspaltung wie aus dem fast parallelen Kurvenverlauf der Protonenkonzentration der CH_3- und NOH-Gruppen zu erkennen ist (Abb. 2), von beiden Gruppen im gleichen Maße erfolgt.

Wie aus Abb. 3 zu ersehen ist, erfolgt die Methanabspaltung linear mit steigender Dosis und wird ab etwa 1 MGy stark beschleunigt. Die Bildung von Äthylen und Äthan ist erst ab etwa 0,15 MGy nachweisbar und steigt mit Zu-

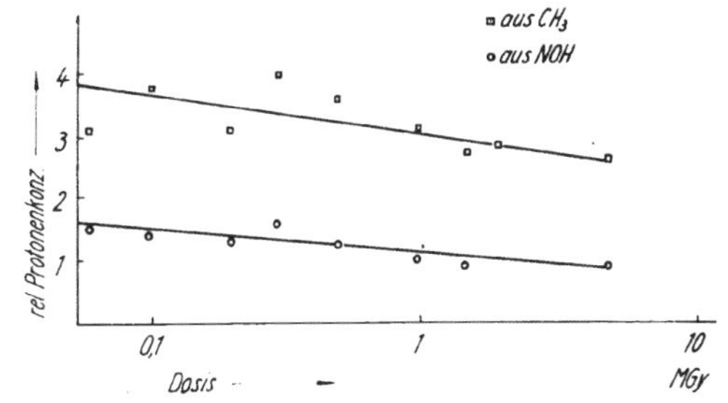

Abb. 2. Abnahme der NMR-Protonensignale der NOH- und CH_3-Gruppen von DMG in Abhängigkeit von der Dosis

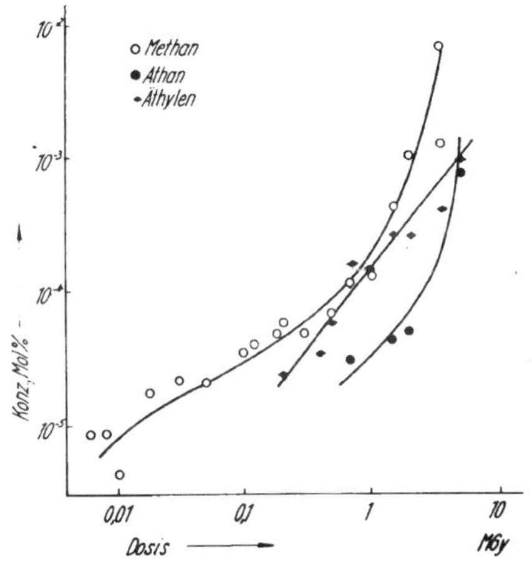

Abb. 3. Gasabspaltung in Abhängigkeit von der Dosis

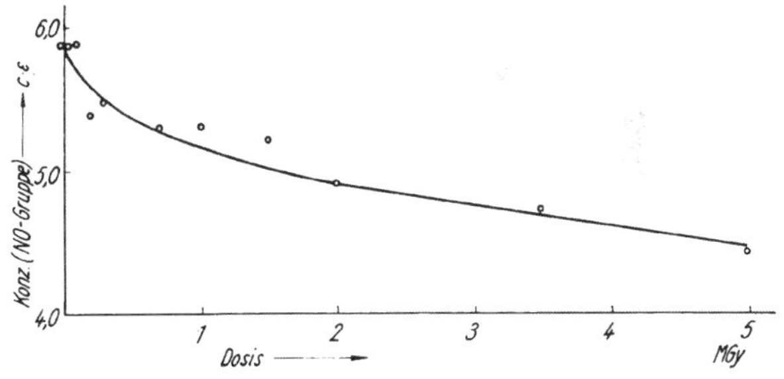

Abb. 4. Abnahme der Konzentration der NO-Gruppe im DMG in Abhängigkeit von Dosis gemessen mit IR-Spektrometer an 2%-igen Lösungen in DMF

nahme der Dosis ebenfalls an. Sie kann sowohl auf einen C-C- und C-N-Bindungsbruch im DMG als auch auf den Zusammenschluß von CH_3-Radikalen zurückgeführt werden.

Der radiolytische Abbau der NO-Gruppe des DMG ist IR-spektroskopisch ab Bestrahlungsdosen > 100 kGy nachweisbar (Abb. 4). Mit zunehmender Dosis nimmt der C-N-Bindungsbruch zu. Bei hohen Dosen von 1 MGy sind etwa 10% und bei 5 MGy etwa 25% der NO-Gruppen abgespalten.

Bei den Fällungsreaktionen wurde aus der Differenz zwischen der theoretischen DMG-Konzentration für den unbestrahlten Komplexbildner und der in Abhängigkeit von der Bestrahlungsdosis ermittelten tatsächlichen Konzentration des fällbaren Reagenz auf den radiolytischen Abbau des DMG geschlossen. In Abb. 5 sind die Ergebnisse zusammengestellt. Sie zeigen in Übereinstimmung mit den Aussagen der NMR und IR-Spektroskopie, daß

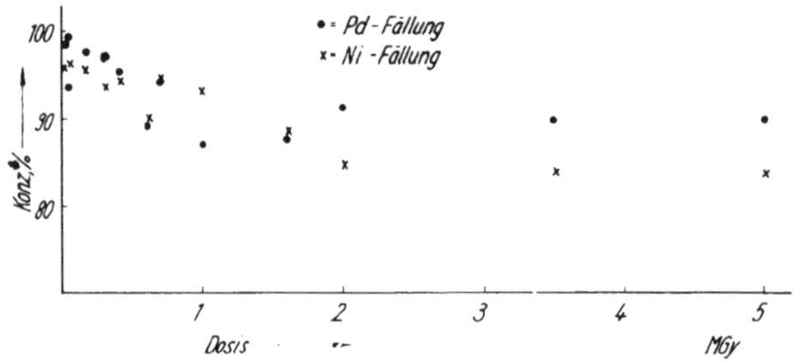

Abb. 5. Abnahme der Fällungskonzentration von DMG in Abhängigkeit von der Dosis

die komplexbildende NOH-Gruppe bis 5 MGy ca. 15% an Wirksamkeit verliert.

Aus der absorbierten Dosis und der Konzentration an Spaltprodukten lassen sich folgende strahlenchemische Ausbeuten berechnen:

$G(H_2) = 0{,}05 \pm 0{,}005$
$G(CH_4) = 0{,}005 \pm 0{,}0025$
$G(C_2H_6) = 0{,}001 \pm 0{,}0005$
$G(C_2H_4) = 0{,}001 \pm 0{,}0005$

Die Radiolyseuntersuchungen des reinen DMG und der Kombination DMG mit Aktivkohle zeigen keine signifikanten Unterschiede. Jedoch ist bei Anwendung von Aktivkohle als Trägermaterial für das DMG auf Lösungsmittelfreiheit zu achten.

Schlußfolgerungen

Die vorliegenden Ergebnisse zum radiolytischen Abbauverhalten von DMG dürften für die Beurteilung von Anwendungsmöglichkeiten des Komplexbildners zur selektiven Abtrennung von Palladium aus Kernbrennstoffrückständen umfassende Aussagen gebracht haben, wie sie bisher unbekannt waren.

Werden Kernbrennstoffe mit einem Abbrand von 25000 MWd/t nach dem Purex-Verfahren wiederaufbereitet, erhält man eine Abfallösung, die nach einer Lagerzeit von 5 Jahren eine Gesamtaktivität von ca. $4 \cdot 10^{12}$ Bq/l hat [15]. Soll aus einer solchen Lösung durch ein selektives Sorptionsverfahren Palladium gewonnen werden, muß man nach einer Abschätzung von *Richter* [15] bei einem Durchsatz von 10 l/h mit einer mittleren Strahlendosisleistung von 480 Gy/h in der Sorptionskolonne rechnen. Bei Einsatz konzentrierter Abfallösungen [17] erreicht die Dosisleistung ca. den zehnfachen Wert.

Die durchgeführten strahlenchemischen Untersuchungen ergaben nun, daß DMG eine relativ strahlenresistente Substanz gegenüber Gammastrahlung im Bereich bis zu Dosen von etwa 100 kGy ist.

Eine weitere Steigerung der Bestrahlungsdosis bis zu 5 MGy führt nur zum teilweisen Abbau der komplexbildenden NOH-Gruppen. Solch hohe Dosen ergeben sich in Sorptionskolonnen mit konzentrierter Abfallösung aber erst nach Standzeiten von etwa 10^3 h.

Bei der Verwendung von mit DMG modifizierten Aktivkohlen für die Pd-Gewinnung aus Kernbrennstoffabfalllösungen wird jedoch nicht an eine mehrmalige Verwendung des modifizierten Sorbens gedacht. Demnach kann man bei solchen Trennversuchen Komplikationen durch eine ungenügende Strahlenresistenz von DMG völlig ausschließen.

Eingegangen am 7. 2. 1983

Literatur

[1] *E. Blasius* u. a., Z. Anal. Chem. **284** (1977) 337
[2] *K. H. Lieser*, u. a., Z. Anal. Chem. **284** (1977) 361
[3] *G. Schmuckler*, Offenlegungsschrift 1964922
[4] *K. Motojima* u. a., Ann. Nucl. Energy **5** (1978) 5
[5] *K. Motijima* u. a., J. Nucl. Sci. and Technol. **16** (1979) 200
[6] *E. S. Palschin* u. a., Ž. anal. Chim. **21** (1966) 954
[7] *B. M. Vanderborght*, *R. E. van Grieken*, Anal. Chem. **49** (1977) 311
[8] *V. V. Skopenko* u. a., Ž. neorg. Chim. **26** (1981) 2477
[9] *D. C. Gregoire*, *A. Chow*, Talanta **22** (1975) 453
[10] *A. M. Gurvic*, *T. B. Gapon*, Zavod. Lab. **23** (1957) 1037
[11] *V. M. Nerezov* u. a., Nauchn. Tr., Vses-Nauchno-Issled Proektn. Inst. Tugoplavkikh Met. Tverd. Splavov **14** (1973) 267
[12] *S. V. Pljuščeva*, *M. M. Senjavin*, Ž. Fiz. Chim. **42** (1968) 2012; **43** (1969) 2150
[13] *R. H. Moore*, US Pt. 3848048 (1974)
[14] *O. E. Jakimočenko*, *Ja. S. Lebedev*, Int. J. Radiat. Phys. Chem. 3/1 (1971) 17–44
[15] *M. Richter*, Isotopenpraxis **17** (1981) 269
[16] *G. O. Müller*, Praktikum der quantitativen chemischen Analyse, S. Hirzel-Verlag, Leipzig 1962
[17] *W. Heinerl*, *P. Grziwa*, Kerntechnik **18** (1976) 272

Isotopenpraxis 19 (1983) 10, S. 345–350

„Gammadefektoskopie" an Kartoffeln mit dem Radionuklid Americium-241

M. Gläser, B. Herold

(Forschungszentrum für Mechanisierung Schlieben/Bornim der Akademie der Landwirtschaftswissenschaften der DDR)[1])

Die zunehmende Industrialisierung der landwirtschaftlichen Produktion bei höheren Qualitätsanforderungen verlangt objektive und hochempfindliche Kontrollverfahren.

Zur zerstörungsfreien Ermittlung innerer Qualitätsmängel an Früchten, z. B. Kartoffelknollen, können lichtoptische Verfahren wegen der starken Streuung nicht umfassend eingesetzt werden. Deshalb wurde unter Einsatz weicher Gammastrahlung (^{241} Am) versucht, „trockene" Hohlstellen, insbesondere bedingt durch Hohlherzigkeit, in Kartoffelknollen zu erkennen. Es konnte gezeigt werden, daß bei einem Erkennungsfehler von etwa 13% diese Mängel (ab ca. 5 mm Durchmesser) zu erkennen sind, auch wenn die Knollen nur in einer durch die Knollenmitte verlaufenden Ebene durchstrahlt werden.

Keywords

americium 241; gamma radiography; nondestructive testing; potatoes; quality control

1. Problemstellung

Zur zerstörungsfreien Ermittlung innerer Qualitätsmängel an Früchten, wie z. B. Kartoffelknollen können Meßverfahren der Durchstrahlung mit Licht eingesetzt werden [1]. Derartige Durchleuchtungsverfahren bieten insbesondere die Möglichkeit, innere Verfärbungen des Fruchtgewebes zu erkennen, sind jedoch zum Nachweis kleiner Defekte (Abmessungen von wenigen Millimetern) weniger geeignet, da die Lichtdurchlässigkeit in hohem Maße von der sehr unterschiedlichen Schichtdicke beeinflußt wird. Im Fruchtgewebe wird Lichtstrahlung außerdem so stark gestreut, daß eine Durchlässigkeitsmessung scharf abgegrenzter Gewebebereiche und damit der Nachweis kleiner Unregelmäßigkeiten ohne Einbeziehung weiterer Meßverfahren praktisch unmöglich ist.

Solche kleinen Defekte können jedoch grundsätzlich mit Hilfe von Meßverfahren der Röntgendurchstrahlung erkannt werden, wobei der quantitative Nachweis von Defekten. z. B. über Bildaufnahmen o. ä. jedoch relativ aufwendig ist [2, 3].

Deshalb sollte die meßtechnische Erfaßbarkeit von kleinen Defekten (Hohlstellen) in Kartoffelknollen näher untersucht werden. Insbesondere sollte für die Defektoskopie an organischen Materialien mit niedriger Dichte die Eignung des langlebigen Radionuklids Americium-241 (Gamma-Strahlungsenergie von 64 keV) geprüft werden.

2. Theoretische Voruntersuchungen

2.1. Bestimmung der optimalen Strahlungsenergie

Wegen des exponentiellen Charakters der Schwächung ionisierender Strahlung durch Stoffschichten ist die Meßempfindlichkeit eine Funktion des Massenschwächungskoeffizienten und damit der Strahlungsenergie. Eine maximale Meßempfindlichkeit liegt dann vor, wenn folgende Bedingung erfüllt ist (z. B. [4, S. 888]):

$$\left(\frac{\mu}{\varrho}\right) \cdot d = 1 \qquad (1)$$

[1]) Anschrift: 1503 Potsdam-Bornim, Max-Eyth-Allee

Hierin bedeuten:

$\frac{\mu}{\varrho}$ = Massenschwächungskoeffizient, im allgemeinen material- und energieabhängig

d = Flächenbelegung, vom Meßproblem vorgegeben

Wegen der außerhalb des optimalen Meßbereiches geringeren Meßempfindlichkeit eignet sich eine bestimmte Strahlungsenergie (Röntgenröhre bestimmter Konstruktion und Betriebsspannung oder Gammastrahlung emittierendes Radionuklid) nur für einen begrenzten Meßbereich.

Unter Verwendung von Gl. (1) kann man nach Vorgabe der Flächenbelegung und des Materials des jeweiligen Meßproblems den günstigsten Massenschwächungskoeffizienten errechnen und die entsprechende Energie ermitteln (z. B. [5, S. 202 und 203]). Für die vorliegenden Untersuchungsobjekte wird als mittlere Schichtdicke der Kartoffeln eine solche von 6 cm angenommen. Das entspricht bei einer Dichte von etwa 1 g cm^{-3} einer maximalen Flächenbelegung von 6 g cm^{-2} (bzw. 60 kg m^{-2}). Damit folgt aus Gl. (1), daß zur Erzielung optimaler Meßempfindlichkeit eine Strahlungsenergie gewählt werden muß, die bei der Transmission durch Kartoffeln einen Massenschwächungskoeffizienten von

$$\left(\frac{\mu}{\varrho}\right)_{\text{optimal}} = 0{,}016 \text{ m}^2 \text{ kg}^{-1}$$

besitzt.

Dieser Massenschwächungskoeffizient liegt vor für vorwiegend Wasser und leichte Elemente enthaltende Substanzen beim Einsatz weicher Röntgen- bzw. Gammastrahlung mit einer Energie um 0,09 MeV. Die genaue Ermittlung unter Berücksichtigung der elementaren Zusammensetzung von Kartoffeln und tabellierter [5, 6] Massenschwächungskoeffizienten für die einzelnen Elemente liefert im in Frage kommenden Energiebereich die in Tab. 1 zusammengestellten Werte. Die theoretischen effektiven Massenschwächungskoeffizienten für Kartoffeln wurden durch anteilige Summierung gemäß Gl. (2) erhalten. Danach wird die weiter oben erfolgte Abschätzung bestätigt. Als Strahlenquellen kommen daher praktisch nur Röntgenröhren bzw. weiche Gammastrahlung emittierende Radionuklide (insbesondere Americium-241 mit einer günstigen Halb-

Tab. 1. Massenschwächungskoeffizienten (nach [3] u. [4], vgl. Abb. 1), Optimalkriterium, Meßempfindlichkeit und Auflösungsvermögen für weiche Röntgen- und Gammastrahlung

Element	Kern-lad.-zahl	Masse-anteil m_i		Massenschwächungskoeffizient $\left(\frac{\mu}{\varrho}\right)$/m² · kg⁻¹ für verschiedene Strahlungsenergien (E_γ)/keV						
		%		25	50	64	75	100	125	150
Wasserstoff	1	1,32		0,035	0,033	0,03226	0,032	0,031	0,027	0,024
Kohlenstoff	6	8,78		0,027	0,017	0,01658	0,015	0,014	0,013	0,0125
Stickstoff	7	0,76		0,033	0,018	0,01720	0,0155	0,014	0,013	0,013
Sauerstoff	8	9,14		0,044	0,021	0,01774	0,016	0,014	0,013	0,013
Wasser	6 (eff.)	80	11,11 (H) 88,99 (O)	0,043	0,022	0,01866	0,018	0,016	0,015	0,014
Kartoffel	theoretisch			0,0415	0,0216	0,01856	0,0177	0,0159	0,0148	0,0142
	experimentell (Abb. 2)					0,01611				
Optimal-kriterium	$\frac{\mu}{\varrho} \cdot d$			2,490	1,296	1,119	1,064	0,954	0,888	0,852
Meßempfind-lichk. n. Gl. (2)				0,206	0,359	0,366	0,367	0,367	0,365	0,363
Auflösungsver-mög. n. Gl. (6)	mm			0,30	0,60	0,68	0,67	0,80	0,86	0,90
$I_0 = 10^5$	%			0,50	1,00	1,14	1,12	1,33	1,43	1,50
$I_0 = 10^4$				0,95	1,90	2,11	2,10	2,46	2,65	2,76
				1,60	3,00	3,52	3,48	4,11	4,42	4,60

wertszeit von 458 a) in Frage.

$$\left(\frac{\mu}{\varrho}\right)^{\text{Kart}} = \frac{1}{100} \sum_{i=O,C,H,N,H_2O} m_i \cdot \left(\frac{\mu}{\varrho}\right)_{E_\gamma i} \tag{2}$$

Für Americium-241-Gammastrahlung wurde der auf diese Weise ermittelte Wert experimentell bestätigt (Abb. 1).

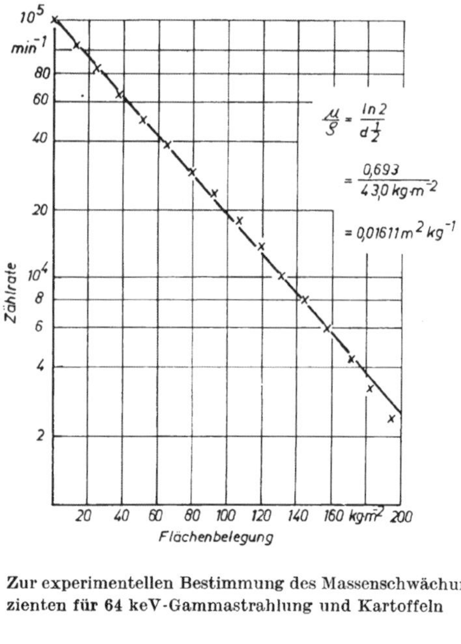

Abb. 1. Zur experimentellen Bestimmung des Massenschwächungskoeffizienten für 64 keV-Gammastrahlung und Kartoffeln

2.2. *Einfluß auf die Erkennbarkeit von Hohlstellen in Kartoffelknollen*

Die Erkennbarkeit von Defekten wird bestimmt durch

1. die Empfindlichkeit des Meßverfahrens für Schichtdickenunterschiede
2. die geometrische Form der Defekte und
3. die natürlich gegebenen Formunregelmäßigkeiten von Kartoffelknollen (Abweichungen von Ellipsoidform, Vorhandensein von Augen usw.).

Die Empfindlichkeit des Meßverfahrens ist zwar der wesentliche Einflußfaktor, jedoch wird dem erreichbaren Auflösungsvermögen bei Messung in einem Medium mit anderen Massenschwächungskoeffizienten als Kartoffelmaterial — in der Regel Luft — dadurch eine Grenze gesetzt, daß das Auftreten von Knollenformunregelmäßigkeiten zur Einführung einer Erkennungsschwelle zur Unterdrükkung dieses Störeinflusses zwingt.

In diesem Zusammenhang wird auch deutlich, daß neben der absoluten Größe der Defekte vor allem ihre geometrische Form, insbesondere der Gradient der Schichtdickenänderung senkrecht zur Durchstrahlungsrichtung die Erkennbarkeit beeinflußt.

3. Experimentelle Untersuchungen

3.1. *Erkennungsgrenzen für Defekte (Hohlstellen)*

Die Wirkung der unter Pkt. 2.2 aufgeführten Einflußfaktoren wurde zunächst an Hand von Röntgenaufnahmen unter optimalen Bedingungen von Knollen mit Defekten

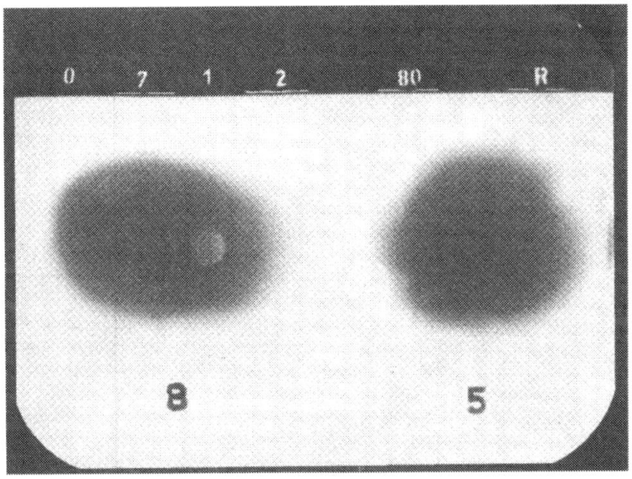

Abb. 2. „Mammographie" von Kartoffelknollen mit eingebrachten definierten Hohlstellen (1:2, vgl. Tab. 2)
8 u. 5: Nummer der Knollen

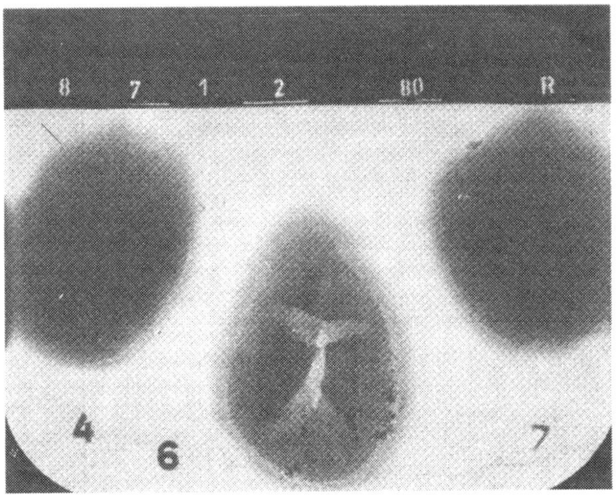

Abb. 3. „Mammographie" von hohlherzigen Kartoffen (1:2, vgl. Tabelle 2)
4, 6 u. 7: Nummer der Knollen

(Hohlstellen) veranschaulicht und eingeschätzt (Abb. 2 und 3 und Tab. 2).

Unter Einsatz einer Mammographie-Röntgenanlage[2]), die eine optimale Detailerkennbarkeit in organischem Gewbe (z. B. 0,1 mm große Mikrokalzifikationen in der weiblichen Brust) ermöglicht, und unter Vorgabe von Hohlstellen bestimmter Ausdehnung in Kartoffelknollen wurde geprüft, bis zu welcher Größenordnung noch Hohlstellen zu erkennen sind. Bei Röhrenspannungen um 30 kV wurden die kontrastreichsten Aufnahmen erzielt (Abb 2 und 3).

Danach sind bei visueller Betrachtung noch zylinderförmige Hohlstellen von 3 mm Durchmesser und 6 mm Höhe mit Sicherheit erkennbar, doppelkegelförmige erst ab 12 mm Durchmesser und 12 mm Höhe (Abb. 2, Tab. 2).

Die Erkennbarkeit von Hohlräumen mit hohen Gradienten der Schichtdickenunterschiede (zylinderförmige Hohlstellen) im Vergleich zu den Randgebieten ist größer. Darüber hinaus sind tiefliegende Augen ein die Erkennungsschwelle ungünstig beeinflussender Faktor.

Trotz dieser Einschränkungen dürfte bei röntgenographischer Untersuchung ein relativ geringer Erkennungsfehler zu erzielen sein. Zur Illustration zeigt Abb. 3 Aufnahmen von 3 Kartoffelknollen mit realen Hohlstellen (Hohlherzigkeit) mit unterschiedlicher Größe und Erkennbarkeit.

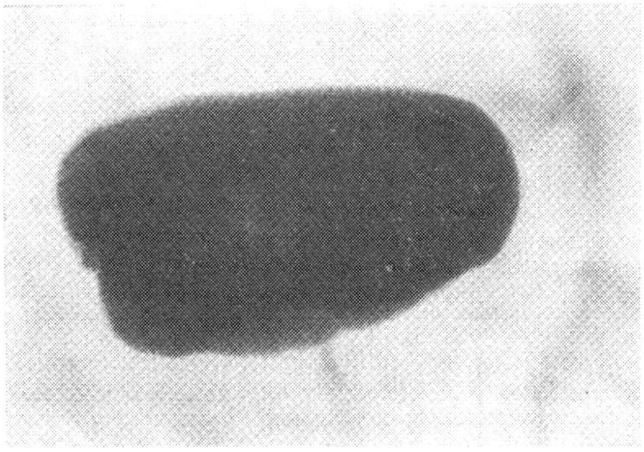

Abb. 4. Gammadefektoskopie einer hohlherzigen Kartoffelknolle

[2]) Dank freundlicher Unerstützung durch die Abteilung Röntgenlabor des VEB Transformatoren- und Röntgenwerk „Herrmann Matern" Dresden, insbesondere durch *R. Luh* und *G. Nagel*

Tab. 2. Röntgenographische Erkennung von Hohlstellen

Knollen-Nr.	Gestalt der Hohlstelle	Durchmesser/mm zu Höhe/mm d. Hohlstelle	Erkennbarkeit	
5	Doppelkegel	a) 3:3	nein	(Abb. 2, 6 u. 7)
		b) 6:6	nein	
		c) 12:12	ja	
8	Zylinder	a) 3:6	ja	
		b) 6:12	ja	
		c) 12:24	ja	
4	langer Spalt	5:6–8	ja	(Abb. 3)
6	Doppelprisma strichmännchenförmig. Aufriß	15:25	ja	
7	Doppelkegel	10:8	ja	
9	Quaderförmig. Aufriß	35:25	ja	(Abb. 8)

3.2. *Versuchsaufbau für die radiometrischen Untersuchungen*

Zur Orientierung wurde zunächst eine Radiographie von einer hohlherzigen Kartoffelknolle unter Einsatz von ^{241}Am-Gammastrahlung angefertigt (Abb. 4). Die Erkennbarkeit erreicht nahezu dieselbe Qualität wie beim Einsatz weicher Röntgenstrahlung, verdeutlicht aber auch die Schwierigkeiten, die durch die Randschichtunregelmäßigkeiten bedingt sind.

Für die Untersuchungen zur Ermittlung von Erkennungsgrenzen und Erkennungsraten unter Einsatz weicher Gammastrahlung wurde folgender Aufbau verwendet:

Eine Americium-241-Strahlungsquelle (Aktivität 3,7 GBq) befand sich in einem Hohlzylinder aus Blei. Durch eine Öffnung (∅ 4 mm) in der der Quelle gegenüber liegenden Stirnfläche trifft der Strahl kollimiert auf die Meßprobe. Diese liegt auf einem kleinen Tischchen zwischen Quelle und Detektor. Zur Erzielung einer hohen Zähleffektivität wird als Detektor ein Szintillationszähler benutzt. Der Szintillator ist gegenüber Streustrahlung durch eine Bleiabschirmung mit Eintrittsöffnung (∅ = 4 mm) für den Nutzstrahl geschützt.

Der Szintillationszähler ist an einen hochstabilen spektrometrischen Meßplatz 22042 angeschlossen. Quelle, Detektor und Probenhalterung sind zwecks Justierung und Einhaltung einer konstanten Meßgeometrie auf einer optischen Bank stationiert. Die Probe kann mittels eines Spindeltriebes quer zur Strahlrichtung vertikal bzw. horizontal bewegt werden, so daß die Möglichkeit einer rasterartigen „Abtastung" der Probe besteht.

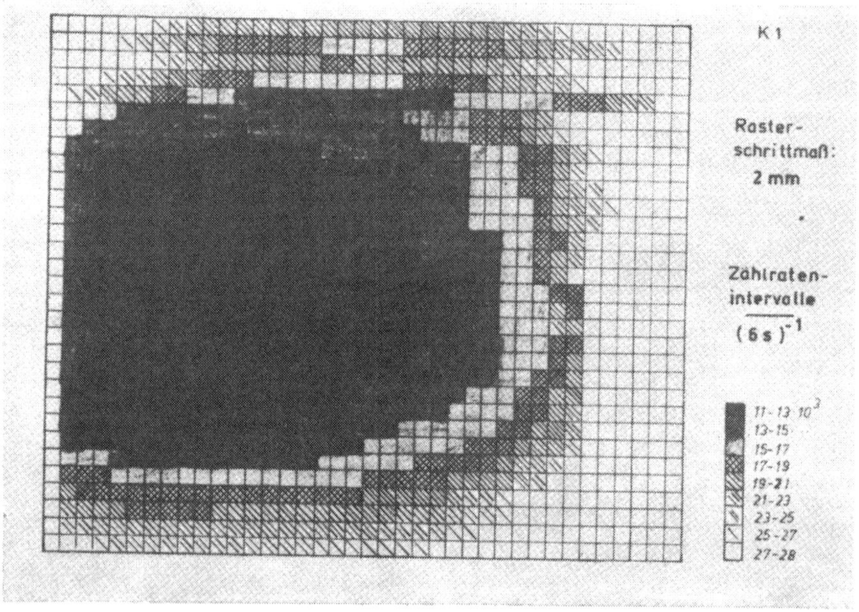

Abb. 5. Rasterartige Durchstrahlung von Kartoffelknollen in Luft mit weicher Gammastrahlung (hohlstellenfreie Knolle)

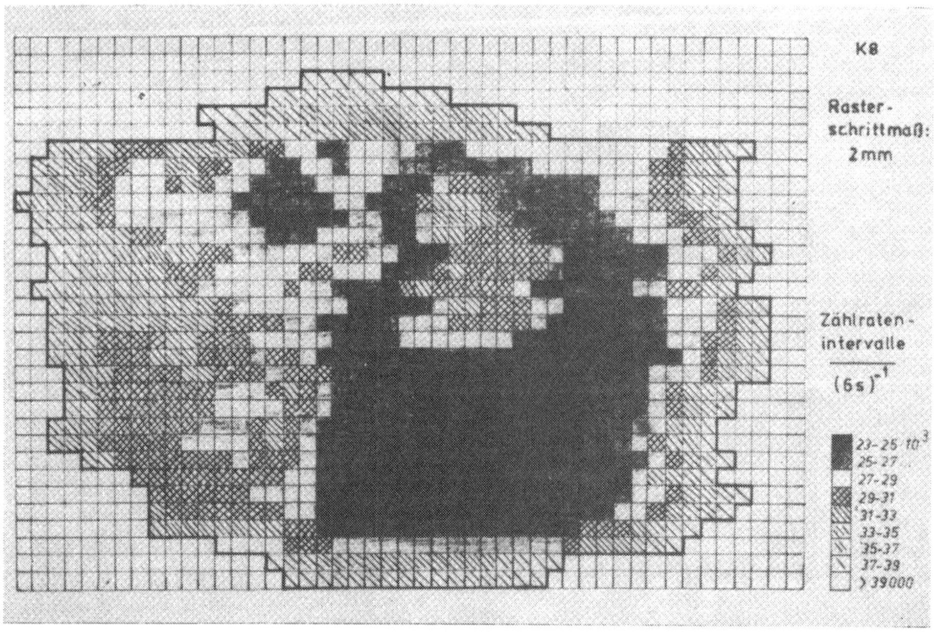

Abb. 6. Rasterartige Durchstrahlung von Kartoffelknollen in Luft mit weicher Gammastrahlung (künstliche, zylinderförmige Hohlstellen, vgl. Abb. 2)

4. Versuchsergebnisse

4.1. *Erkennungsgrenze*

Zur Ermittlung einer unteren Erkennungsgrenze wurden die mit definierten Hohlstellen versehenen Knollen einer Raster-Durchstrahlung unterworfen (Abb. 5 bis 8). Abb. 5 zeigt zunächst das Raster-Durchstrahlungsbild einer gesunden Knolle. Dann folgen die Durchstrahlungsergebnisse (Abb. 6 bis 8) an denselben Knollen, die bereits für die röntenographischen Aufnahmen verwendet wurden (Abb. 2 und 3). Berücksichtigt man das Auftreten von natürlichen Formunregelmäßigkeiten der Knollen, so können nach den vorliegenden Untersuchungen folgende Erkennungsgrenzen angegeben werden:

zylinderförmige Hohlestellen ab 5 mm ∅
und 5 mm Höhe
doppelkegelförmige Hohlstellen ab 15 mm ∅
und 15 mm Höhe.

Dabei wurde der Schwellwert für die Entscheidung, ob ein Defekt vorliegt, im Abstand der zweifachen Standardabweichung vom zu erwartenden Zählratenwert für hohlstellenfreie Knollen festgelegt.

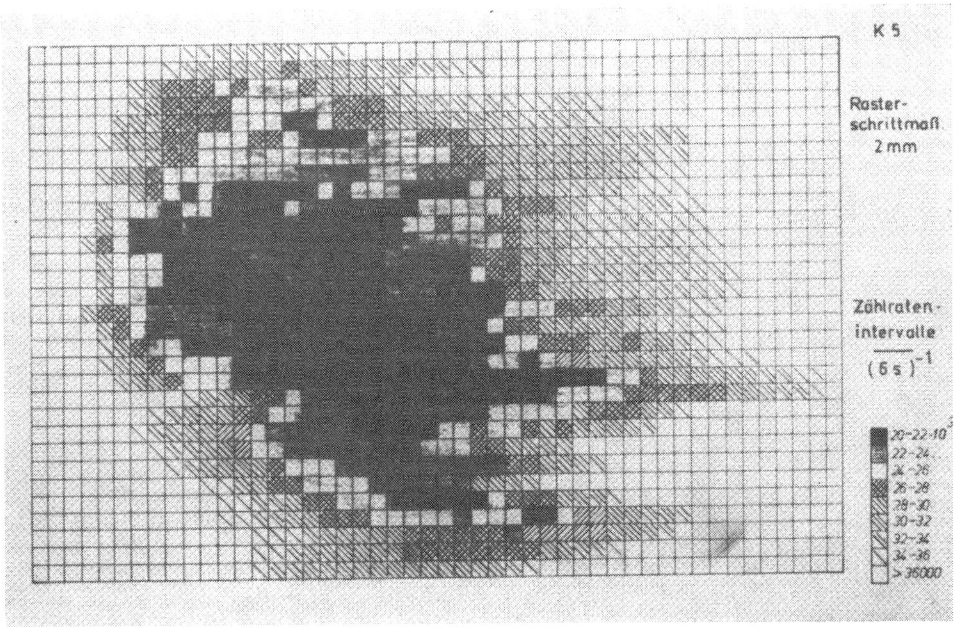

Abb. 7. Rasterartige Durchstrahlung von Kartoffelknollen in Luft mit weicher Gammastrahlung (künstliche, doppelkegelförmige Hohlstellen, vgl. Abb. 3)

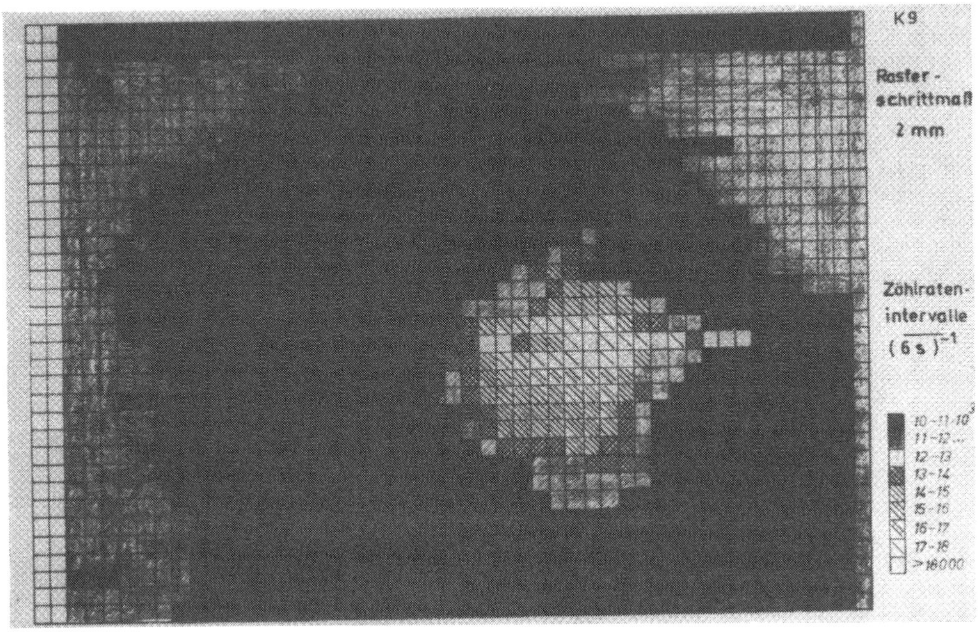

Abb. 8. Rasterartige Durchstrahlung von Kartoffelknollen in Wassereinbettung mit weicher Gammastrahlung (natürliche Hohlstelle; Hohlherzigkeit)

4.2. *Erkennungsfehler bei Messung in Luft*

Zur Ermittlung des Erkennungsfehlers, der ein Anhaltspunkt für die Praktikabilität eines Verfahren ist, wurden ca. 50 Kartoffelknollen, unter denen ein relativ hoher Anteil mit Hohlstellen sein sollte, radiometrisch untersucht.

Tab. 3. Erkennungsfehler für hohlherzige Kartoffeln bei radiometrischer Durchstrahlung in einer Mittelebene

Gesamtanzahl	darunter Knollen mit Hohlstellen	Anzahl radiometr. „erkannter" Hohlstellen	darunter vorgetäuschte, nicht reale Hohlstellen	Erkennungsfehler %
In Luft 88	53	47	1	13
In Wasser 47	24	20	1	21

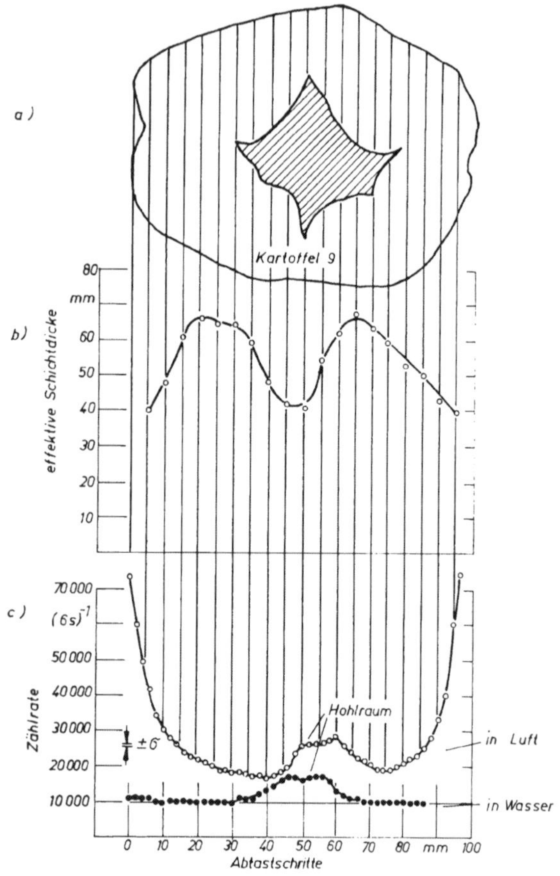

Abb. 9. Durchstrahlung von Kartoffelknollen in ihrer Mittelebene mit weicher Gammastrahlung (vgl. Abb. 8)

a) Schnitt in Durchstrahlungsebene; b) Effektive Schichtdicke quer zur Durchstrahlungsebene; c) Zählratenverlauf quer über die Durchstrahlungsebene (in Luft bzw. Wasser „eingebettete Knolle")

Dabei wurden — bereits vereinfachend — die Knollen nicht über ihren gesamten Querschnitt rasterartig abgetastet, sondern nur in einer durch ihre Mitte verlaufenden Ebene. Die Abtastung erfolgte in 2-mm-Schritten. Der kreisförmige Strahlquerschnitt betrug ca. 12,5 mm^2. Die Knollen wurden in Luft durchgemessen. Die Ergebnisse sind in Tab. 3 zusammengestellt. Ein Meßbeispiel zeigt Abb. 9.

4.3. *Erkennungsfehler bei Messungen in einem Medium mit „angepaßtem" Massenschwächungskoeffizienten*

Angeregt durch einen Literaturhinweis [2] wurde geprüft, ob durch „Einbettung" der Knolle in ein Medium mit „angepaßtem" Massenschwächungskoeffizient eine Verringerung des Erkennungsfehlers möglich ist. Zu diesem Zweck wurden die Knollen während der Messung in ein planparalleles, mit Wasser gefülltes, Gefäß gebracht. Der Massenschwächungskoeffizient von reinem Wasser unterscheidet sich nur um weniger als 1% von demjenigen von Kartoffelknollen [6] (s. auch Tab. 1). Durch die praktizierte „Einbettung" sollten insbesondere Erkennungsfehler durch Oberflächenunregelmäßigkeiten (Augen, Einbuchtungen) weitgehend vermieden werden (Abb. 9).

Die Ergebnisse (Tab. 3) bestätigen diese Überlegungen nicht. Über die Ursachen dafür (evtl. teilweise Auffüllung der Hohlstellen durch eindringendes Wasser) ist derzeit noch keine eindeutige Aussage möglich.

5. Schlußbemerkungen

Es konnte gezeigt werden, daß unter Einsatz weicher Gammastrahlung einer Energie von 64 keV (^{241}Am) in Kartoffelknollen vorhandene „trockene" Hohlstellen, insbesondere bedingt durch Hohlherzigkeit, bei einem Erkennungsfehler von etwa 13% nachgewiesen werden können, auch wenn die Knollen nur in einer durch die Knollenmitte verlaufenden Ebene durchstrahlt werden. Unter Berücksichtigung der naturgegebenen Knollenform sind bei einem Strahlquerschnitt von 12,5 mm^2 praktisch zylinderförmige Hohlstellen ab 5 mm Durchmesser und 5 mm Höhe bzw. kegelförmige Hohlstellen ab 15 mm Durchmesser und 15 mm Höhe mit Sicherheit zu erkennen. Eine Verbesserung kann durch Ausmessung der Kartoffelknollen in verschiedenen Ebenen erfolgen, evtl. durch Einsatz eines aus mehreren Einzeldetektoren bestehenden Liniendetektors, an dem die Knollen vorbeigeführt werden.

Eingegangen am 23. 12. 1982
Überarbeitete Fassung am 2. 3. 1983

Literatur

[1] *G. S. Birth*, American Potato Journal **37** (1960) 53
[2] *E. F. Finney* jr., *K. H. Norris*, American Potato Journal **50** (1973) 1
[3] *R. G. Diener, J. P. Mitchell, M. L. Rhoten*, Agricultural Engineering **51** (1970) 356
[4] *W. Hartmann* (Hrsg.), Meßverfahren unter Anwendung ionisierender Strahlung. Leipzig: Akademische Verlagsgesellschaft Geest und Portig K. G. 1969
[5] *K. Wirtz, K. H. Beckurts*, Elementare Neutronenphysik. Berlin, Göttingen, Heidelberg: Springer-Verlag 1956
[6] *R. G. Jaeger* (Hrsg.) u. a., Engineering compendium on radiation shielding, Vol. I. Berlin, Heidelberg, New York: Springer-Verlag 1968

Isotopenpraxis **19** (1983) 10, S. 351 – 352

Kurze Originalmitteilungen

Untersuchungen zum Adsorptions-Desorptions-Verhalten von $^{99}MoO_4^{--}$ an Al_2O_3 in Abhängigkeit von der Temperatur

G. Bernhard

(Akademie der Wissenschaften der DDR, Zentralinstitut für Kernforschung Rossendorf)[1])

Keywords

adsorption, aluminiumoxid, desorption, molybdenum isotopes, uranium

In der Anlage zur Molybdänproduktion Rossendorf (AMOR) werden aluminiumgemantelte Brennelemente des Rossendorfer Forschungsreaktors zum Zwecke der ^{99}Mo-Separation chemisch aufgearbeitet. Die Abtrennung der ^{99}Mo-Aktivität aus der Spaltproduktlösung erfolgt mittels eines Batch-Verfahrens durch Adsorption an Al_2O_3. Nach einem Waschprozeß wird das $^{99}Mo_4^{--}$ mit 1 N NH_3-Lösung vom Al_2O_3 desorbiert.

Ausgangspunkt der Untersuchungen zur Temperaturabhängigkeit der Adsorption – Desorption von $^{99}MoO_4^{--}$ an Al_2O_3 war die Erkenntnis, daß mit steigender Temperatur der Brennelementlösung bei der Adsorption, der in 1 N NH_3 desorbierbare Anteil der ^{99}Mo-Aktivität sinkt.

Versuchsdurchführung

In einer Kapsolutflasche werden zu 10 ml der zu untersuchenden Lösung 1,6 g Al_2O_3 (VEB Chemiewerk Greiz-Dölau, chloridfrei gewaschen) gegeben. Die Brennelementlösung ist 1,6 N an $Al(NO_3)_3$, 0,025 N $UO_2(NO_3)_2$ und enthält ca. 0,2 N freie HNO_3 sowie feindisperses Si, SiO_2 (ca. 0,25 g/l). Zur Adsorption wird die thermostatisierte Kapsolutflasche 30 min geschüttelt. Danach wird der Flascheninhalt in eine Generatorsäule entleert und das Al_2O_3 mit 5 ml 0,1 N HNO_3, 5 ml H_2O und 5 ml 0,01 N NH_3-Lösung gewaschen. Anschließend wird mit 10 ml 1 N NH_3-Lösung, erwärmt auf die entsprechende Desorptionstemperatur, das $^{99}MoO_4^{--}$ eluiert. Von diesen Lösungen werden Meßproben abgenommen.

Die Berechnung des Verteilungskoeffizienten der Adsorption K_{d_A} erfolgt nach der Beziehung:

$$K_{d_A} = \frac{A_a \cdot V_{Lsg.}}{A_{Lsg.} \cdot m_a}$$

A_a = an Al_2O_3 adsorbiertes ^{99}Mo(Bq)
A_{Lsg} = Restgehalt des ^{99}Mo in der Lösung
V_{Lsg} = Lösungsvolumen, ml
m_a = Al_2O_3, g

Der Verteilungskoeffizient der Desorption K_{d_D} errechnet sich zu:

$$K_{d_D} = \frac{A_E \cdot m_a}{A_a \cdot V_E}$$

A_E = Aktivität ^{99}Mo im Eluat, Bq
V_E = Eluatvolumen, ml

Die Ausbeute G und $^{99}MoO_4^{--}$ nach dem Separationsverfahren wird ermittelt nach:

$$G = \frac{A_G}{A_E} \cdot 100\ (\%)$$

A_G = Aktivität der Aufgabelösung, Bq

Ergebnisse

In Abb. 1 ist der Verteilungskoeffizient der Adsorption von $^{99}MoO_4^{--}$ an Al_2O_3 in 0,1 N HNO_3 (Kurve 1) und in Brennelementlösung (Kurve 2, Restsäure 0,14 N; Kurve 3, Restsäure 0,71 N) in Abhängigkeit von der Temperatur dargestellt. Es ist zu erkennen, daß der K_{d_A}-Wert mit steigender Temperatur zunimmt. Diese Zunahme ist stark säurekonzentrationsabhängig und vergrößert sich mit dieser. Wird die bei unterschiedlichen Temperaturen an Al_2O_3 adsorbierte $^{99}MoO_4^{--}$-Aktivität bei einer Temperatur von 25 °C desorbiert, ergibt sich die in Abb. 2 dargestellte Abhängigkeit des Verteilungskoeffizienten der Desorption von der Adsorptionstemperatur. Es zeigt sich, daß durch Erhöhung der Adsorptionstemperatur der K_{d_D}-Wert deut-

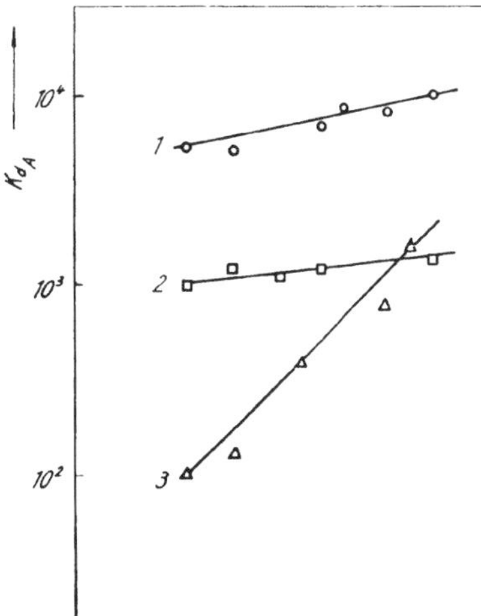

Abb. 1. Abhängigkeit der $^{99}MoO_4^{--}$-Adsorption an Al_2O_3 von der Temperatur

1 – 0,1 N HNO_3; 2 – Brennelementlösung 0,14 N HNO_3; 3 – Brennelementlösung 0,71 N HNO_3

[1]) Anschrift: 8051 Dresden, PF 19

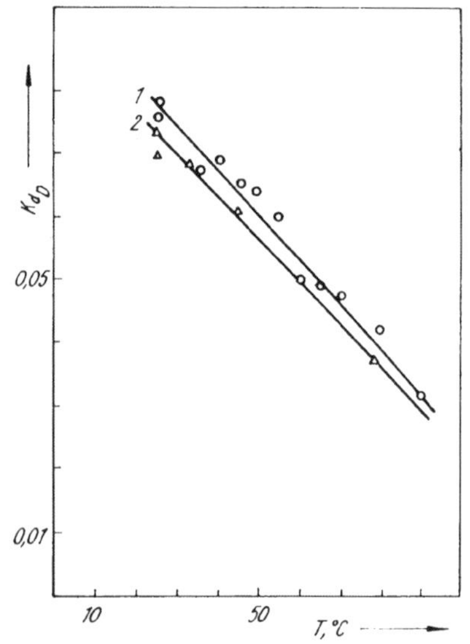

Abb. 2. Abhängigkeit der $^{99}MoO_4^{--}$-Desorption bei 25 °C vom Al_2O_3 von der Adsorptionstemperatur

1 — 0,1 N HNO_3; 2 — Brennelementlösung 0,14 N HNO_3

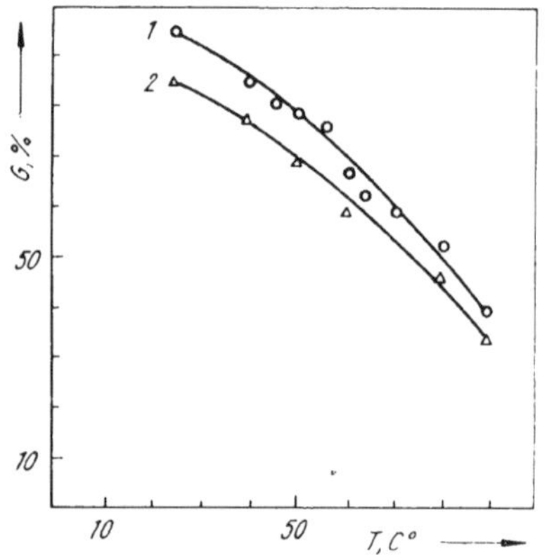

Abb. 3. Abhängigkeit der $^{99}MoO_4^{--}$-Ausbeute von der Adsorptionstemperatur (Desorptionstemperatur = 25 °C)

1 — 0,1 N HNO_3; 2 — Brennelementlösung 0,14 N HNO_3

lich sinkt. Die Unterschiede der K_{dD}-Werte in Abhängigkeit von der Acidität bei der Adsorption sind bedeutend geringer. Wird das $^{99}MoO_4^{--}$ bei 25 °C aus 0,1 N HNO_3 an Al_2O_3 adsorbiert und anschließend mit 1 N NH_3 bei 25 °C desorbiert, beträgt der K_{dD}-Wert 0,078, aus 1,5 N HNO_3 0,62.

Für ein Produktionsverfahren zur Gewinnung von $^{99}MoO_4^{--}$ durch chromatographische Trennung an Al_3O_3 ist es übersichtlich, die Abhängigkeit der Ausbeute (Verhältnis der vorhandenen zur abgetrennten Aktivität) von beeinflussenden Parametern darzustellen. In Abb. 3 ist die Ausbeute (G) an $^{99}MoO_4^{--}$ in Abhängigkeit von der Adsorptionstemperatur bei einer Desorptionstemperatur von 25 °C angegeben. Aus dieser Abhängigkeit ist abzuleiten, daß die Adsorption von $^{99}MoO_4^{--}$ an Al_2O_3 bei 25 °C erfolgen sollte, wenn die Desorption bei Normaltemperatur durchgeführt wird. Das bedeutet, daß die Kernbrennstoff-Spaltproduktlösung die nach der Auflösung des Brennelementes eine Temperatur von 108 °C hat, abgekühlt werden muß, bevor Al_2O_3 zur Adsorption des $^{99}MoO_4^{--}$ zugegeben wird. Wird dieser Temperatureinfluß nicht beachtet, sind Ausbeuteverminderungen bis zu 50% möglich. Die Ausbeuteverluste können auch vermindert werden, wenn die Desorptionstemperatur gleich der Adsorptionstemperatur ist. Betragen sowohl die Adsorptions- als auch die Desorptionstemperatur bei der $^{99}MoO_4^{--}$-Separation aus einer Brennelementlösung 60 °C, werden 75% der vorhandenen Aktivität im Eluat erhalten. Ist die Desorptionstemperatur (siehe Abb. 3, Kurve 2) 25 °C, ist die $^{99}MoO_4^{--}$-Ausbeute nur 58%.

Diskussion

Bekannt sind die Abhängigkeiten des Verteilungskoeffizienten der Adsorption von $^{99}MoO_4^{--}$ an Al_2O_3 von der Säurenormalität [1], von der Uranylnitrat- und Aluminiumnitrat-Konzentration der Lösungen [2]. *Boeßert* u. *Reichel* [3] untersuchten die Abhängigkeit der Spaltmolybdänadsorption vom Chloridgehalt des Al_2O_3 und von der Kontaktzeit. Dabei zeigt sich, daß der Adsorptions-Verteilungskoeffizient des $^{99}MoO_4^{--}$ mit zunehmendem Chloridgehalt steigt. Bei anschließender Desorption ist ein gegenläufiges Verhalten erkennbar. Untersuchungen zum Adsorptionsverhalten von $^{99}MoO_4^{--}$ an Al_2O_3 in Abhängigkeit von der Temperatur zeigen eine Erhöhung des Verteilungskoeffizienten mit steigender Temperatur. Diese Steigerung ist durch Chemisorptionsprozesse bedingt, da reine Adsorptionsprozesse einen konträren Temperaturverlauf zeigen. Dieser chemisorbierte Anteil des $^{99}MoO_4^{--}$ wird bei Desorptionstemperaturen, die geringer sind als die der Adsorptionstemperatur nur zum Teil desorbiert. Der Anteil des $^{99}MoO_4^{--}$ im Eluat wird durch eine Erhöhung der Desorptionstemperatur (Desorptionstemperatur $\geqq$ Adsorptionstemperatur) gesteigert, ohne bei gleichen Kontaktzeiten für Adsorption und Desorption die Ausbeuten an $^{99}MoO_4^{--}$ im Eluat zu erreichen, die bei einer Adsorption und Desorption bei 25 °C erreicht werden. Der Temperatureffekt kann bei der Abtrennung des $^{99}MoO_4^{--}$, vor allem aus Kernbrennstoff-Spaltproduktlösungen höherer Acidität, ausgenutzt werden. Geschieht die Abtrennung zur Gewinnung des $^{99}MoO_4^{--}$, wird demgemäß die Aktivität vom Ionenaustauscher desorbiert, sind bei der Adsorption Temperaturen $<$ 30 °C einzuhalten.

Eingegangen am 2. 2. 1983

Literatur

[1] *S. El Bayoumi, O. Hladik, R. Münze*, Isotopenpraxis **9** (1973) 131
[2] *K. Jantsch*, ZfK-294 (1975) 69
[3] *W. Boeßert, B. Reichel*, ZfK-393 (1973) 82

Isotopenpraxis 19 (1983) 10, S. 353–358

Informationen

IAEA. Der Generaldirektor der Internationalen Atomenergie-Organisation (IAEA) hat folgende Ernennungen bekanntgegeben:

– Dr. *Peter M. Tempus* (Schweiz) wird Stellvertretender Generaldirektor an der Hauptabteilung für Sicherheitskontrolle.
– *Volkmar Schuricht* (Deutsche Demokratische Republik) wird Direktor der Inspektionsabteilung B an der Hauptabteilung für Sicherheitskontrolle.
– *Leslie Thorne* (Großbritanien) wurde zu einem leitenden Angestellten im Direktorenrang an der Hauptabteilung für Sicherheitskontrolle befördert.
– *Svein Thorstensen* (Norwegen) wurde zu einem leitenden Angestellten im Direktorenrang an der Hauptabteilung für Sicherheitskontrolle befördert.

(IAEA-Pressekommuniqué)

IAEA. Gemeinsam mit der Weltgesundheitsorganisation (WHO) veranstaltete die Internationale Atomenergie-Organisation (AEA) ein Internationales Symposium über die biologischen Wirkungen schwacher Strahlungen im April 1983 in Venedig. Strahlungen sind von jeher ein Bestandteil unserer natürlichen Umwelt, und eine Strahlenbelastung der Bevölkerung ist zum größten Teil unvermeidlich. Andererseits werden nukleare Vefahren auf den verschiedensten Gebieten, beispielsweise in der industriellen Radiographie (zur Qualitätskontrolle) oder in der Medizin (zur Diagnose und Behandlung) immer mehr angewandt, und der Anteil der Kernenergie an der Elektrizitätsversorgung nimmt ständig zu: Gegenwärtig liegt er bei 10% der Weltproduktion. Durch den Einsatz nuklearer Verfahren wird die Strahlenbelastung bestimmter Berufsgruppen und der Bevölkerung im allgemeinen unweigerlich erhöht. Die Gefährdung des einzelnen und die „Schädigung" der Bevölkerung durch Strahlungen wurden von innerstaatlichen und internationalen Expertengremien begutachtet, und heute herrscht verbreitet die Meinung, daß bei Einhaltung der von der Internationalen Strahlenschutzkommission (ICRP, ein 1928 gegründetes unabhängiges nichtstaatliches Expertengremium, hat die Aufgabe, Empfehlungen für den Schutz vor ionisierenden Strahlen auszuarbeiten. Ihre Mitglieder werden auf Grund ihrer persönlichen Qualifikation in den Bereichen Röntgenologie, Strahlenschutzphysik, Genetik und auf verwandten Gebieten gewählt. Die Empfehlungen der ICRP werden seit 50 Jahren von den für Strahlenschutz zuständigen innerstaatlichen und internationalen Organen allgemein akzeptiert.) empfohlenen Höchstdosen keine wesentlichen radiologischen Wirkungen zu erwarten sind. Trotzdem werden in den letzten Jahrzehnten die Wirkungen geringer Dosen ionisierender Strahlen auf den Menschen und seine Umwelt in der Öffentlichkeit und in wissenschaftlichen Kreisen immer stärker diskutiert. Dabei richtet sich die Aufmerksamkeit der Öffentlichkeit vor allem auf die potentiellen Risiken, die durch das vorgesehene Wachstum der nuklearen Elektrizitätsgewinnung und die damit verbundene Freisetzung zunehmender Mengen radioaktiv kontaminierter Abwässer entstehen können. Deshalb hat die Kernindustrie, um diese radioaktive „Verseuchung" zu verhüten, Maßnahmen und Verfahren entwickelt, die im allgemeinen als fortschrittlicher und verfeinerter gelten als die gegen andere industrielle Schadstoffe angewandten. Die Kernenergiegewinnung trägt somit nur ganz wenig zur Strahlenbelastung der Bevölkerung bei – in den Industrieländern etwas über 0,1% der durchschnittlichen Individualdosis.

Auf dem Symposion in Venedig werden Informationen über die Beurteilung der biologischen Wirkungen schwacher Strahlungen ausgetauscht und die jüngsten Fortschritte erörtert.

Zwei Symposien ähnlicher Art haben die IAEA und die WHO bereits 1975 und 1978 veranstaltet. Seither wurden zahlreiche experimentelle Daten gesammelt und neue Methoden und Verfahren entwickelt. Insbesondere wurden die höchstzulässigen Dosen revidiert.

(IAEA-Presseinformation)

IAEA. Die bis Ende 1982 von der Internationalen Atomenergie-Organisation (IAEA) gesammelten Daten zeigen, daß der Anteil der Kernenergie an der Elektrizitätsproduktion der Welt stetig wächst – obwohl in einigen Ländern Kernkraftwerksprojekte aufgegeben wurden und trotz der gegenwärtigen weltweiten Rezession.

Ende des Jahres waren in 25 Ländern 294 Leistungsreaktoren mit einer Gesamtkapazität von 173.108 MWe im Rahmen des Verbundnetzes in Betrieb; 251 mit einer Gesamtkapazität von 197.860 MWe waren in 27 Ländern im Bau; und weitere 156 mit einer Gesamtkapazität von 152.605 MWe waren in der Planung. Während des Jahres wurden 21 Anlagen an das Verbundnetz angeschlossen, wodurch die nukleare Weltkapazität um 17.812 MWe anstieg. Ferner wurden sieben neue Kraftwerke mit insgesamt 6.693 MWe in Auftrag gegeben (vier in Frankreich und drei in Japan). Andererseits wurden während des Jahres auch 18 bereits erteilte Kraftwerksbestellungen storniert – zwei in Italien und die restlichen in den USA.

Aus den der IAEA übermittelten vorläufigen Daten geht hervor, daß Finnland, was den Anteil der Kernenergie an der Stromproduktion betrifft, mit 40,3% derzeit an der Spitze steht, Frankreich kommt mit 38,7% an zweiter Stelle. Diese Reihung wird sich aber voraussichtlich rasch ändern, vor allem dadurch, daß die gegenwärtig im Bau befindlichen oder kurz vor der Inbetriebnahme stehenden Anlagen in Betrieb genommen werden. Ende 1981 standen auf Grund bestätigter Daten folgende Länder an der Spitze: 1. Frankreich (37,7%), 2. Finnland (35,8%), 3. Schweden (35,3%), 4. Schweiz (28,1%), 5. Belgien (25,3%), 6. Bulgarien (24,7%), 7. Japan (17,3%), 8. Bundesrepublik Deutschland (14,6%), 9. Großbritannien (12,7%) und 10. USA (11,9%). In den Ländern der Europäischen Wirtschaftsgemeinschaft stieg der Anteil der Kernenergie während des Jahres 1982 durch Neuinbetriebnahmen insgesamt von 16,7% auf 18,8% an.

In den USA, wo im letzten Jahrzehnt die Stornierungen meist zahlreicher waren als die neuen Aufträge, rechnet man trotzdem damit, daß die Kernenergie, sobald die Daten für 1982 zur Verfügung stehen, als Energieträger nach der

Kohle an zweiter Stelle stehen wird. Im allgemeinen betrachten die Industrieländer — die eine gewisse energiewirtschaftliche Unabhängigkeit und stärkere Vielfalt ihrer Energieversorgung anstreben — den Einsatz der Kernenergie zur Elektrizitätsgewinnung als wirtschaftlich unbedingt vorteilhaft. Auf der Konferenz über Erfahrungen mit der Kenergie, die Ende letzten Jahres in Wien stattfand, äußerte *W. Kenneth Davis*, vom Department of Energy der USA, in seinem Referat die verbreite Ansicht, daß „in den Vereinigten Staaten zunehmender Bedarf an elektrischer Energie *und* Kernenergie besteht. Obwohl wir noch dabei sind, uns auf die neuen Realitäten des Wirtschaftswachstums und des Energieverbrauchs einzustellen, können wir (die Kernindustrie) ganz besonders zur Befriedigung dieses Bedarfs beitragen." *Rémy Carle*, von Electricité de France, erklärte in einem Vortrag, den er im November in Washington, D.C., beim Atomic Industrial Forum hielt, den Grund für das umfangreiche und weiter wachsende Kernenergieprogramm seines Landes: Obwohl der Ölpreis in den letzten Jahren auf das Vierfache angestiegen ist, „zeigen unsere Prognosen, daß es dank der zunehmenden Einbeziehung der Kernenergie in die Energiewirtschaft möglich sein wird, eine Preispolitik mit konstanten Preisen in wertbeständigen Francs zu verfolgen." (IAEA-Pressemitteilung)

IAEA. Anläßlich der Internationalen Konferenz über Erfahrungen mit der Kernenergiegewinnung, die von der IAEA (Internationalen Atomenergie-Organisation) vom 13. bis 17. 9. 82 in Wien abgehalten wurde, nahm auch der damalige Vorsitzende der britischen Atomenergiebehörde United Kingdom Atomic Energy Authority (UKAEA), Sir *Walter Marshall*, zu den möglichen Folgen höchst unwahrscheinlicher schwerer Kernenergieunfälle Stellung. Dabei machte er den Vorschlag, die voraussichtlichen langfristigen Folgen solcher Unfälle — kürzere Lebenserwartung, höhere Wahrscheinlichkeit tödlicher Krebserkrankungen — den entsprechenden Folgen der Nikotinsucht gegenüberzustellen. Als Beispiel führte er einen „gewaltigen" Kernenergieunfall an, durch den 10 Mio. Einwohner von London eine Dosis von je 1 rem (10 mSv) (die natürliche Strahlenbelastung beträgt pro Jahr ca. 15% dieser Dosis) empfangen würden. Die langfristige Folge davon (bis zu 30 Jahre), berechnet aufgrund der Richtlinien der Internationalen Kommission für Strahlenschutz (ICRP-26), wären 1250 Todesfälle. Dieser langfristige Schaden wäre rechnerisch der gleiche, wenn jeder Einwohner von London jeden Sonntag eine zwanzigstel Zigarette rauchen würde.

(Atomwirtschaft/Atomtechnik — 4/83)

Welt. Das United Nations Scientific Committee on the Effects of Atomic Radiation (UNSCEAR) hat seinen 8. Bericht über die Strahlenbelastung der Bevölkerung für das Jahr 1982 veröffentlicht.

Zur natürlichen Strahlenbelastung wird ausgeführt, daß infolge der stark gewachsenen Zahl von Messungen die Werte der externen Bestrahlung sehr viel genauer als früher ermittelt werden. Für die äußere und die innere Bestrahlung werden in diesem Bereich die Werte erstmals als effektive Äquivalentdosis angegeben. Es wird hervorgehoben, daß etwa die Hälfte der effektiven Äquivalentdosis aus natürlichen Strahlenquellen durch Radon in der Luft in Gebäuden hervorgerufen wird. Zur Strahlenbelastung durch die Medizin wird ausgeführt, daß Durchschnittswerte in sich wenig sinnvoll sind, da sie aus einer großen Spanne der Strahlenbelastung ermittelt werden, die zwischen Null beim unbehandelten Menschen bis zum Zehntausendfachen der natürlichen Strahlenbelastung infolge Radiotherapie variiert. Auch hier gestattet die Verwendung der effektiven Äquivalentdosis eine klarere Aussage als früher. In den Industrieländern ergibt sich aus diagnostischen Methoden eine medizinische Strahlenbelastung der Bevölkerung, die etwa bei der Hälfte der natürlichen liegt. Der Beitrag der therapeutischen Strahlenbelastung wurde nicht abgeschätzt.

Der Beitrag der Kernkraftwerke hat sich von 0,0001% im Jahr 1960 auf ca. 1,01% des natürlichen Untergrundes 1980 erhöht. Diese Steigerung ist direkt proportional zum Anstieg der friedlichen Nutzung der Kernenergie. Auch hier variieren die Belastungswerte sehr stark mit der Entfernung von der Kernenergieanlage. Typische Werte am Zaun der Kernkraftwerke lagen zwischen Bruchteilen von 1% bis zu wenigen Prozent der effektiven Äquivalentdosis der natürlichen Strahlenbelastung. Der Langzeiteffekt der Kernenergienutzung wurde für einen Zeitraum von 500 Jahren abgeschätzt. Für eine Kernenergieerzeugung auf dem Niveau von 1980 entspräche der Einfluß der abgegebenen Radioaktivität als Langzeiteffekt auf die Bevölkerung etwa zwei Stunden der natürlichen Strahlenbelastung, während die Kurzzeit-Komponente auf etwa 30 Minuten geschätzt wird. (Atomwirtschaft/Atomtechnik — März 1983

Welt. Nach einer Übersicht der International Maritime Organization (IMO) fanden 1982 Versenkungsaktionen radioaktiver Abfälle der Länder Belgien, Großbritannien, Niederlande und Schweiz statt. Die Versenkungsaktion für Großbritannien umfaßte 2697 t radioaktiver Abfälle in 2833 Behältern mit insgesamt 1264 Curie Alpha-Radioaktivität, 38129 Curie Beta/Gamma ohne Tritium und 63383 Curie Tritium. Diese Aktion wurde am 16. 8. 82 beendet.

Die gemeinsame belgisch-niederländisch-schweizerische Aktion wurde am 6. 9. 82 durchgeführt. Sie umfaßte 8996 t radioaktiver Abfälle in 11797 Behältern, davon 100 Curie Alpha-Radioaktivität, 64 Curie Ra-226, 11410 Curie Beta/Gamma ohne Tritium und 14066 Curie Tritium.

(Atomwirtschaft/Atomtechnik — 4/83)

Europa. Als einen Durchbruch in der physikalischen Grundlagenforschung und den größten Erfolg seit der Einrichtung von CERN im Jahre 1954 bezeichnete CERN-Generaldirektor Prof. *H. Schopper* die Entdeckung des „W intermediären Vektorbosons" durch zwei Arbeitsgruppen des Europäischen Laboratoriums für Teilchenphysik. In den Arbeitsgruppen UA-1 und UA-2 arbeiteten insgesamt 180 Physiker aus acht europäischen Ländern und den USA seit fünf Jahren zusammen.

Die Entdeckung des W Bosons wurde durch den Umbau des 450-GeV-Protonensynchrotrons (SPS) möglich, das jetzt auch als Proton-Antiproton-Kollider mit je 270 GeV eingesetzt werden kann. Die im Speicherringbetrieb erreichbare Gesamtenergie von 540 GeV reicht aus, das äußerst schwere „W intermediäre Vektorboson" zu erzeugen, dessen Masse einer Energie von 80 GeV entspricht. Das UA-1-Team entdeckte in einer Gesamtheit von ca. 1. Mrd. Kollisionen fünf Ereignisse, die das geladene W Boson erkennen ließen, die UA-2-Gruppe bei gleichfalls 1 Mrd. Kollisionen vier Ereignisse. Bosonen gelten als Überträgerteilchen der sogenannten schwachen Kraft, die für den Beta-Zerfall und für die Energieproduktion in der Sonne verantwortlich ist. Nach der neueren Theorie beweisen die Bosonen, die nach dem englischen Ausdruck für schwach zur Gruppe der Weakonen gezählt werden, den Zusammenhang zwischen der elektromagnetischen Kraft und der schwachen Kraft. Die Wissenschaftler *A. Salam*, *S. Glashow* und *S. Weinberg* hatten die Weakonen theoretisch vorhergesagt und für ihre „elektroschwache Theorie" 1979 den Physik-Nobelpreis erhalten. Mit den Versuchsergebnissen bei CERN wurde diese Theorie jetzt bestätigt.

(Atomwirtschaft/Atomtechnik – März 1983)

DDR. Im Zentralinstitut für Isotopen- und Strahlenforschung (ZfI) der Akademie der Wissenschaften der DDR wurde eine Studie über die „Nutzung von Magnetbanddiensten zur Informationsversorgung der Physikwissenschaften" als Sonder-Nr. 1 des ZfI-Mitteilungen (146 S.) angefertigt. In ihr werden eine Reihe von Problemen behandelt, die für die unmittelbare Arbeit mit physikrelevanten Magnetbanddiensten wie ASSISTENT, INIS und INSPEC in Informationseinrichtungen verschiedener Größe von Bedeutung sind. Ausgehend von der Tatsache, daß die Bedeutung der Datenverarbeitung als einer der wesentlichsten Intensivierungs- und Rationalisierungsfaktoren in der wissenschaftlich-technischen Information stark zunimmt und daß die Magnetbanddienste bei der nationalen und internationalen Kooperation auf diesem Gebiet zu den z. Z. wichtigsten Informationsmitteln zählen, werden in der Studie Inhalt, Semantik, Syntax sowie programmtechnische und technische Voraussetzungen beschrieben.

Einen wesentlichen Platz in der Studie nehmen die unmittelbaren technologischen Probleme und Arbeitsprozesse bei der Nutzung von Magnetbanddiensten ein, bis hin zu Beispielen für Nutzeranleitungen.

Die Studie erscheint als Sondernummer und ist erhältlich gegen eine Schutzgebühr (IAP) von 8,20 M für das gedruckte Heft bzw. 6,80 M als Mikrofiche. Bestellungen sind zu richten an:

Wissenschaftliche Redaktion ZIDIS
AdW – ZfI – IRZ
DDR-7010 Leipzig
Petersstraße 26.

BRD. Im Heft 2 der Atomkernernergie/Kerntechnik sind u. a. folgende Beiträge enthalten:

Nettoenergiebilanz von Tokamak-Fusions-Kraftwerken (*R. Bünde*);
Anhäufungseffekte bei der elastischen Neutronenstreuung an Bor-Isotopen (*S. Komoda* u. *T. Sekiya*);
Messungen der totalen Neutronen-Wirkungsquerschnitte für Germanium bei Energien unter 2 eV (*M. Salama*);
Aluminium-induzierte verstärkte Verfärbung einiger Boratgläser durch γ-Bestrahlung (*A. M. Nassar, M. M. Morsi* u. *M. A. Adawi*);
Die innere Bremsstrahlung bei β-Strahlen mit erlaubten Übergängen (*S. El-Konsol, A. M. Basha, A. S. Gaafar, A. A. El-Sayed* u. *A. S. Yousef*);
Einige Diskussionen über relativistische Bahnformmodifikationen bei den Planeten (*Kl.-H. Maull*, Gegenbemerkung und eine Ergänzung durch *E. Bagge*);

(Atomkerntechnik/Kerntechnik 42 (2) (1983) 73–144)

BRD. Vom Kernforschungszentrum Karlsruhe (KfK) wurde ein neues Trainingsgerät entwickelt, mit dem die Bewältigung von Krisensituationen in Kernkraftwerken durch das verantwortliche Personal erprobt werden kann. Es handelt sich um ein Datenverarbeitungssystem, basierend auf einem preiswerten und leicht zu handhabenden Mini-Büro-Computer. Neben einem Bildschirmplatz für die Ablaufsteuerung stehen vier Bildschirmgeräte und zwei Drucker für die Darstellung des Störfallablaufs bereit.

Dem System werden aufbereitete reale Störfallprotokolle oder erdachte Ereignisse eingegeben, besonders solche, die von den üblichen Kraftwerkssimulatoren nicht erfaßt werden können. Eine eingehende Untersuchung des KfK-Institutes für Reaktorentwicklung zur Frage des menschlichen Verhaltens bei Krisensituationen in Kernkraftwerken ergab, daß die Warten der deutschen Anlagen in ergonomischer Hinsicht erheblich besser ausgelegt sind, als die Warten mancher amerikanischer Anlagen. Auch das Training des Schichtpersonals habe einen hohen Stand erreicht. Dennoch sei es wünschenswert, daß insbesondere das Führungspersonal auch auf extrem unwahrscheinliche, hypothetische Situationen eingestellt sei.

(Atomwirtschaft/Atomtechnik – März 1983)

BRD. Nach der Erweiterung des UNILAC-Beschleunigers, der jetzt die doppelte Teilchenenergie und bis zu zehnmal höhere Intensitäten für schwere Ionen liefert, eröffnen sich bei der Gesellschaft für Schwerionen (GSI) verbesserte Forschungsmöglichkeiten.

Versuche zur Atomphysik und zur Kernphysik können durch die höhere Intensität in kürzerer Zeit oder bei gleicher Zeit mit besserem Ergebnis durchgeführt werden. Vor allem bei der Erzeugung von bisher unbekannten Atomkernen müssen immer seltenere Isotope beschleunigt und mit dem geeigneten Zielkern verschmolzen werden. So war z. B. die Verbesserung des Beschleunigers eine wichtige Voraussetzung zur Entdeckung des Elementes 109, das durch Verschmelzung von Eisen-58-Ionen mit Wismut-209 erzeugt wurde. Durch die Verdopplung der Teilchenenergie auf max. 20 MeV können jetzt höhere Ladungszustände bei den Ionen erzeugt und untersucht werden.

(Atomwirtschaft/Atomtechnik – März 1983)

BRD. Die Zeitschrift Atomwirtschaft/Atomtechnik hat im März-Heft 1983 als Schwerpunktthema die Brennelemente.

Unter anderen sind folgende Artikel dazu enthalten:
R. Holzer, H. Stehle: LWR-Brennelemente für erhöhten Abbrand;
W. Böhm: Stand und Entwicklungstendenzen der BE-Einsatzplanung in Leichtwasserreaktoren;
R. Schwarzwälder, K. Hnilica, W. Fehrenbach, H. Hoffmann: Zur Methodik der BE-Einsatzplanung in Leichtwasserreaktoren;
H.-P. Bochmann, A. Matting: Genehmigungsverfahren für Brennelement-Fabriken;
H. Roepenack: MOX-Brennstäbe für Schnelle Brüter und Leichtwasserreaktoren;
K. G. Hackstein, M. Hrovat: Brennelemente niedriger U-235-Anreicherung für Forschungs- und Materialtestreaktoren;
G. Bäro, F. D'Annucci, W. Stindt: Inspektion und Reparatur von Brennelementen;
Fröhlich: Wasserchemie und Korrosionsprobleme bei Kernreaktoren (Bericht über das Internationale IAEA-Symposium in Wien);
St. Halaszovich, R. Odoj: Entsorgung zwischen Technologie und Gesellschaftspolitik (Bericht über die DAtF-Informationstagung in Bonn).

Außerdem ist in diesem Heft die Schnellstatistik der Kernkraftwerke 1982 – Weltübersicht – enthalten.

(Atomwirtschaft/Atomtechnik 28 (1983) No. 3)

BRD. Das Brennelement-Forschungslabor der Kraftwerk Union AG (KWU) und die Fertigungsentwicklung der Reaktor-Brennelement Union (RBU) haben gemeinsam in fünfjähriger Arbeit ein neues Verfahren zur Herstellung von Kernnbrennstofftabletten entwickelt. Diese Tabletten werden in sogenannte Brennstäbe, dünne Rohre aus einer Zirkoniumlegierung, eingefüllt, aus den Brennstäben ihrerseits werden Brennelemente hergestellt. Die neue Produktionsmethode für die Brennstofftabletten (Pellets) bietet die Möglichkeit, den Kernbrennstoff für unterschiedliche Kundenanforderungen variabel herzustellen, wobei die im Brennstoff Uran enthaltene Energie noch besser als bisher ausgenutzt werden kann. Insbesondere gelingt es durch das neue Verfahren, die Glühtemperatur zur Verdichtung (Sintern) der aus Uranoxid-Pulver gepreßten Pellets von 1700 °C auf 1100 °C zu senken und die Glühzeit auf weniger als die Hälfte herabzusetzen. Bei dem neuen Prozeß wird außerdem nur noch einfach zu handhabendes Kohlendioxidgas benötigt, das wirtschaftlicher ist als das bisher verwendete Kohlenwasserstoffgas.

(Atomwirtschaft/Atomtechnik – 4/83)

BRD. Anfang März wurde mit den vorbereitenden Arbeiten für abschließende Tests und die Verfüllung der Tiefbohrung Gorleben 1002 durch die Deutsche Gesellschaft zum Bau und Betrieb von Endlagern für Abfallstoffe (DBE) begonnen. Anschließend wird die Tiefbohrung Gorleben 1003 abschließend getestet und verfüllt. Außerdem ist die Teilverfüllung der Schachtvorbohrungen Gorleben 5001 und 5002 vorgesehen. (Atomwirtschaft/Atomtechnik – 4/83)

BRD. Im Rahmen eines vom Bundesminister des Innern geförderten Forschungsvorhabens wird beim TÜV Rheinland versucht, aus der amtlichen Todesursachenstatistik Schlüsse aus der regional unterschiedlichen Meldehäufigkeit von bestimmten Krebserkrankungen als Todesursaache zu ziehen. Regional unterschiedliche Umweltfaktoren, z. B. die teilweise recht unterschiedliche natürliche Umgebungsstrahlung, könnten damit auf ihren möglichen Einfluß auf die Krebstodesrate geprüft werden.

Als erste Ergebnisse dieses Forschungsvorhabens erschienen im Verlag TÜV Rheinland die ersten beiden vom BMI herausgegebenen Bände einer Veröffentlichungsreihe „Die Krebssterblichkeit in der Bundesrepublik Deutschland 1970 bis 1978 — Ergebnisse des Informationssystems über Krebsmortalität und Krebscharakteristika (IKK)". Der von den statistischen Landesämtern zur Verfügung gestellte Datenbestand enthält z. Z. die Zahl der Sterbefälle nach Todesursache, Bevölkerungsstand, Grundstrahlung und Flächennutzung. Die Basisdaten des IKK entstammen der amtlichen Todesursachenstatistik und der Wohnbevölkerungsstatistik der Bundesländer und Berlin (West) für die Berichtsjahre 1970—78. Die in den ersten beiden Bänden vorgelegten Auswertungen stellen die altersstandardisierten Sterberaten für 25 verschiedene Todesarten vor, davon 23 verschiedene Krebslokalisationen. Mit dem Gesamtwerk des IKK erhält die Fachwelt ein Instrument, um vielfach behauptete Zusammenhänge zwischen Krebssterblichkeit und Umwelteinflüssen zu analysieren.

(Atomwirtschaft/Atomtechnik – 4/83)

BRD. Im Heft 4 der Zeitschrift Atomwirtschaft/Atomtechnik **28** (1983) sind u. a. folgende Beiträge enthalten:
Bewertung der Verfügbarkeit von Kernkraftwerken (*U. Hansen*);
Das Wasserstoffproblem (*H. Karwat*);
Druckprüfung des Reaktorbehälters des THTR-300;
Zur ergänzenden Darstellung der Ergebnisse von Risikostudien (*J. Ehrhardt, A. Bayer*);
Einfluß der DWR-Umwälzschleifen auf den Blowdown (LOBI-Projekt) (*W. L. Riebold, P. Mörk-Mörkenstein, L. Piplies, H. Städtke*);
Neue Kernkraftwerke in der BRD 1983 (Teil I des Berichtes über Bauvorhaben und Projekte in Europa).

ČSSR. Das Forschungsinstitut für die Produktion und Anwendung von Radioisotopen (ÚVVVR — Institute for Research, Production and Application of Radioisotopes) in Prag hat einen neuen Katalog für Strahlenquellen herausgebracht. Außerdem ist eine Ergänzungsliste für markierte Verbindungen (^{3}H- und ^{14}C-markierte Nukleotide, cADP u. a.) erschienen.

ÚVVVR — 10227 Prag 10, Radiovà 1

UdSSR. „Atomnaje Energija" **54** (1983) 1 enthält u. a. folgende Beiträge (*russisch*):
My years with I. V. Kurchatov — In Memory of M. D. Millionshchikov (*A. P. Alexandrov*);
Transfer and Deposition of Radioactive Isotopes in a Convectional Flow of Sodium (*A. S. Zhilkin* u. a.);
Induced Radioactivity of Certain Compositions of Concrete under Irradiation Conditions in the 680-MeV Proton Accelerator (*V. F. Kasyanov* u. a.);
The Radiation Situation at the 16-MeV Microtron Electron Accelerator (*A. G. Belov* u. a.);
Electromagnetic Mass-Separator for Radioisotopes (*R. I. Lyubtsev* u. a.);
Determination of the Coefficient of Isotope Separation During Chemical Exchange by the Multistage Compression Method (*S. D. Moiseyev* u. a.);
A Device for Determining the Direction of Subterranean Flow of Water Dislocated in a Borehole (*I. G. Skovorodnikov*);
The Effect of Boring Concrete on Decreasing Induced Activity Under Conditions of Proton Irradiation (*V. F. Kasyanov* u. a.);
Plastic Scintillators for Registering Fast Neutrons (*D. V. Viktorov* u. a.);
Increasing the Efficiency of Using Neutron Flux in a Neutron-Activation Installation with Central Location of a Californium Source (*B. S. Vakhtin* u. a.);
Additional Radiation Factors on High-Current Electron Accelerators (*L. F. Belovodski* u. a.);
Taking Account of Background Self-Neutron Radiation when Determining the Composition of the Mixture of Fissioning Nuclides from the Delayed Neutrons (*B. P. Maksyutenko* u. a.);
Using a Plastic Scintillator Counter for Measuring High-Energy Neutron Spectra (*V. E. Aleinikov* u. a.).

Großbritannien. In letzter Zeit sind folgende Reports der United Kingdom Energy Authority (UKAEA) erschienen:

AERE-M 3252 A gamma doserate surveillance system. By *A. T. Rolls*, *G. F. Snelling* and *F. D. Seymour*. July 1982. 13 pp.

AERE-R 10032 Random number generators for the normal and gamma distributions using the ratio of uniforms method. By *I. Robertson* and *L. A. Walls*. December, 1980. 58 pp.

AERE-R 10470 Determination of rare earth elements in geological materials by neutron activation analysis using a group separation and high resolution gamma-ray spectrometry. By *T. W. Sanders* and *S. J. Wright*. September, 1982. 21 pp.

AERE-R 10554 The simultaneous measurement of particle size, velocity and mass transfer in a pulsed two-phase flow field. By *M. L. Yeoman, N. S. Lightfoot* and *A. P. Morse*. June, 1982. 21 pp.

AERE-M 3258 A beta-in-air monitor surveillance, alarm and record system. By *G. F. Snelling* and *F. D. Seymour*. August, 1982. 26 pp.

AERE-M 3261 A process plant cooling water monitor for fission product breakthrough. By *A. T. Rolls, G. F. Snelling* and *F. D. Seymour*. August, 1982. 17 pp.

AERE-R 9066 (rev) Initial experience of manufacture and property measurements on piezoelectric cable. By *B. H. Broomfield*. November, 1981, 23 pp.

AERE-R 10496 Selection of low-activity elements for inclusion in structural materials for fusion reactors. By *O. N. Jarvis*. June, 1982. 46 pp.

AERE-R 10631 Computer modelling studies of photochemical air pollution formation in power station plumes in the United Kingdom. By *R. W. Derwent*. August, 1982. 54 pp.

ND-R 461(S) The Auger spectra of some ternary lithium compounds. By *I. F. Ferguson, D. R. Masters* and *M. Turek*. September, 1982. 19 pp.

ND-R 803(W) A comparison of measured fission gas releases for Windscale AGR fuel irradiated above 18 GWd/tU with those predicted using the computer code MINIPAT D. By *N. Beatham, R. Hargreaves, R. D. Walker* and *R. Kirkbridge*. August, 1982. 26 pp. (Atom 314 – 1982)

AERE-R 9835 A programmable and remotely controlled laser scanning system. By *M. L. Yeman, H. Bowes* and *A. Taylor*. July, 1982. 21 pp.

AERE-R 10533 MA27 – A set of Fortran subroutines for solving sparse symmetric sets of linear equations. By *I. S. Duff* and *J. K. Reid*. July, 1982. 44 pp.

AERE-R 10574 An assessment using thin layer activation of a corrosion inhibitor treatment. By *J. Asher, J. W. Webb, N. J. M. Wilkins* and *P. F. Lawrence*. August, 1982, 13 pp.

AERE-R 10587 The measurement of particle size, velocity and concentration with a non-invasive technique in multi-phase flow fields. By *M. L. Yeoma*. July, 1982. 42 pp.

AERE-R 10597 A review of the U-238 resolved resonance parameters. By *M. C. Moxon* and *M. G. Sowerby*. August, 1982. 36 pp.

AERE-R 10598 Permeation of gases through solids. 2. An assessment of measurements of the steady state permeability of H and its isotopes through Fe, Fe based alloys and some commercial steels. By *A. D. LeClaire*. July, 1982. 31 pp.

AERE-R 10627 Accurate ultrasonic time delay measurement with EMAT's for residual stress determination. By *D. R. Allen*. October, 1982. 22 pp.

ND-R 804(S) Fatigue crack growth in simulated weld heat-affected zones in A533-B steel. By *A. McMinn*. October, 1982, 29 pp.

AERE-R 10629 The measurement of ultrasonic phase delay for residual stress determination. By *D. R. Allen* and *W. H. B. Cooper*. October, 1982. 36 pp.

AERE-R 10737 Testing of hazardous wastes to assess their suitability of landfill disposal. By *P. J. Young* and *D. C. Wilson*. November, 1982. 18 pp.

CLM-R225 Electron cyclotron resonance heating at Culham. Contributions to the ECRH workshop, Brusselts, June 1982. By *W. H. M. Clark, J. Hugill, C. N. Lashmore-Davies, M. O'Brian, A. C. Riviere, D. C. Robinson, D. F. H. Start, R. A. Cairns* and *J. Owen*. May, 1982. 34 pp.

CLM-R228 The thermal stability and equilibrium of peripheral plasmas. By *D. E. T. F. Ashby* and *M. H. Hughes*. September, 1982. 24 pp.

ETSU-R 13 (Vol. 1) Strategic review of the renewable energy technologies. An economic assessment, Vol. 1. Executive summary and main report. November, 1982. 110 pp.

ETSU-R 13 (Vol. 2) Strategic review of the renewable energy technologies. An economic assessment, Vol. 2. Appendices containing detailed analyses. November, 1982. 380 pp.

ND-R 720(W) Fission-gas releases during post-irradiation annealing of UO_2 fuel from WAGR pins of burnup up to 5500 MWd/te U. By *D. G. Bridge, D. Cordall* and *J. Young*. July, 1982. 48 pp.

ND-R 848(S) Technical review of UK work on environmentally assisted cracking in PWR pressure vessel materials. By *B. Tomkins*. November, 1982. 30 pp.

ND-R 852(R) Assessment of UKOSRP crack growth data to investigate the remaining life of offshore structures following inspection. By *A. M. Clayton*. November, 1982. 57 pp.

The following reports are available from the Editorial Section, Safety and Reliability Directorate, UKAEA, Wigshaw Lane, Culcheth, Warrington WA3 4NE:

SRD-R238 The transition from pressure-energy to gravity-driven slumping during catastrophic release of a denser-than-air gas. By *S. F. Jagger*. December, 1982. 12 pp.

SRD-R246 Seismic design criteria and their applicability to major hazard plant within the United Kingdom. By *M. H. A. G. Alderson*. December, 1982. 63 pp.

(Atom 316 – 1983)

Großbritannien. Der National Radiological Protection Board (NRPB) hat den Bericht NRPB-R135 „An assessment of the Radiological Impact of the Windscale Reactor Fire, October 1957" veröffentlicht. Dieser Bericht beschreibt einführend den Unfallhergang und anschließend die verschiedenen Verfahren der Messung von Verteilung und Radioaktivität der abgegebenen Radionuklide sowie die notwendigen Rechenmethoden, ihre Fehlergrenzen und schließlich die errechneten Gesundheitseffekte.

Die Kollektivdosis wurde aus verschiedenen Messungen der Umweltradioaktivität in England, Wales und in Nordeuropa errechnet. Meßwerte standen aus der Luft, aus Milch, Gras und Oberflächenablagerungen zur Verfügung. Wegen der während des Unfalls am 10. und 11. 10. 57 herrschenden komplizierten meteorologischen Bedingungen waren die Verteilungsrechnungen sehr komplex. Die gesamte Schilddrüsendosis der britischen Bevölkerung wurde zu $2,5 \cdot 10^4$ Mann-Sievert (2,5 Mio. Mann-rem) ermittelt und entspricht etwa 95% der jährlichen kollektiven Schilddrüsendosis in Europa. Die Untersuchung ergab, daß Jod-131 fast für die gesamte Kollektiv-Schilddrüsendosis und für einen großen Teil der Strahlenbelastung allgemein verantwortlich war. Caesium-137 trug zur Gesamtdosis hauptsächlich durch von Bodenablagerungen ausgehende externe Bestrahlung und über den Ingestionspfad bei.

Als Risikokoeffizient über die Lebenszeit der Bevölkerung in der Nähe der Windscale-Anlagen wurde mit konservativen Randbedingungen der Faktor $1 \cdot 10^{-3}$ eingesetzt. Das so ermittelte Risiko basiert auf den höchsten gemessenen Schilddrüsendosen von Kindern, die in der Größenordnung von 100 mSv lagen. Die höchsten Dosisbelastungen bei Erwachsenen lagen um den Faktor 5 darunter.

Unter Zugrundelegung der gesamten Schilddrüsendosis der britischen Bevölkerung in Höhe von $2,5 \cdot 10^4$ Mann-Sievert ergeben sich aus dem Windscale-Unfall rechnerisch 260 Schilddrüsen-Krebsfälle zusätzlich zu den sonstigen Schilddrüsen-Krebsfällen, von denen 13 tödlich sein könnten. Wenn die Standard-Lebenserwartung nach Beginn der Bestrahlung bei der gegebenen Altersverteilung durchschnittlich 40 Jahre betragen würde, könnten jährlich 6,5 zusätzliche Schilddrüsen-Krebsfälle auftreten. In England und Wales wurden 1968 insgesamt 666 neue Schilddrüsen-Krebsfälle registriert, eine Rate, die in den folgenden Jahren ziemlich konstant blieb.

(Atomwirtschaft/Atomtechnik März 1983)

Großbritannien. Die British Nuclear Fuels Ltd. (BNFL) hat sich mit ihrer Belegschaft und den zuständigen Gewerkschaften über die Prozedur zur Anerkennung von eventuell durch berufliche Belastung ausgelöste Krebsfälle geeinigt. Diese Einigung läßt weiterhin die Möglichkeit einer Anerkennung auf dem Prozeßwege offen. In diesem Falle geht es jedoch nur um Anerkennung oder Abweisung der Klage, während die betriebliche Vereinbarung vorsieht, auch geringere Strahlenbelastungen auf freiwilliger Basis im Falle einer Krebserkrankung als Ursache oder Mitursache gelten zu lassen.

Die Höhe einer Ausgleichszahlung wird von einem unabhängigen Expertengremium unter Vorsitz von Lord *Gregson of Stockport* festgestellt.

(Atomwirtschaft/Atomtechnik – 4/83)

Niederlande. Nach einer Übersicht des niederländischen Ministers für Raumordnung und Umweltfragen fielen in den Jahren 1978 bis 1982 im Bereich des Energieonderzoek Centrum Nederland (ECN) im Jahresmittel 2850 Fässer mit je 200 l behandelter radioaktiver Abfälle an. Die Verarbeitungskapazität im ECN beträgt 3000 Fässer/Jahr. Im gleichen Zeitraum wurden auch radioaktive Abfälle im Atlantischen Ozean versenkt. Es waren 1978 insgesamt 1534 Curie/714 m³, 1979 846 Curie/981 m³, 1980 539 Curie/876 m³, 1981 1864 Curie/943 m³, 1982 1497 Curie/1364 m³.

(Atomwirtschaft/Atomtechnik – 4/83)

Japan. In dem schwerwassermoderierten, leichtwassergekühlten 165-WMe-Prototypreaktor Fugen wurde erstmals in Japan gewonnenes Uran, das auch in Japan angereichert wurde, als Brennstoff eingesetzt: Beim Brennstoffwechsel am 29. 12. 82 wurden drei neue Brennelemente mit japanischem Brennstoff geladen.

(Atomwirtschaft/Atomtechnik – 4/83)

Isotopenpraxis **19** (1983) 10, S. 358–364

Buchbesprechungen

Der Teil und das Ganze
Gespräche im Umkreis der Atomphysik

W. Heisenberg. 5. Aufl. München – Zürich: R. Piper & Co Verlag 1981, 334 S.; geb. DM 25,–

Mag die Auffassung *Thomas Carlyle's*, daß die Weltgeschichte nichts anderes sei als eine Biographie großer Männer, eine ins heroenhafte führende Überbetonung des einzelnen sein, in der geschichtlichen Entwicklung der Naturwissenschaften haben Einzelpersönlichkeiten einen sehr hohen und fördernden Anteil. Sie sind es, welche in die kaum überschaubaren Ansammlungen von Beobachtungen, Daten, Fakten und Detailzusammenhänge ordnend eingreifen, eine höhere Zusammenschau erlauben und die Abstraktion verdinglichen, greifbar machen. *Werner Heisenberg* war eine solche Persönlichkeit, und es ist ungemein anregend, seine Lebens- und Entdeckergeschichte lesen zu können, seine Erfahrungen und Begegnungen auf diesem Wege verfolgen zu können; ja es ist sogar „die Absicht des Verfassers, auch dem der modernen Atomphysik Fernstehenden einen Eindruck von den Denkbewegungen zu vermitteln, die die Entstehungsgeschichte dieser Wissenschaft begleitet haben".

Das Buch beginnt mit den Jahren nach dem 1. Weltkrieg, mit dem ersten Erstaunen Heisenbergs über die bildhafte Vorstellung, daß die Atome in einem Molekül durch Haken und Ösen zusammengehalten würden, die damals angebotene anthropomorphe Form einer chemischen Bindung. Dem Studium bei *Arnold Sommerfeld* in München folgen seine schöpferischen Jahre, die der Quantenmechanik gewidmet waren, aufs engste verbunden mit der Verbreitung des Kopenhagener Geistes der Quantentheorie, der Schule des Altmeisters *Niels Bohr*. In diesen 20er Jahren ist immerhin so viel an anwendungsbereitem Wissen auf dem Gebiet der Atomphysik angehäuft worden, daß *Heisenberg* 1929 in Leipzig auf einem Festvortrag sagen konnte, diese Art von Physik sei im Grunde langweilig und jegliches weitere Interesse daran würde erlöschen. Dabei waren zu diesem Zeitpunkt weder das Neutron, noch das Neutrino oder die Kernspaltung entdeckt. Tatsache ist nur, daß die Arbeiten von *Pauli, Jordan* und *Heisenberg* die Quantentheorie der Wellenfelder zu einem gewissen Abschluß und Stillstand geführt haben.

Das Erstaunliche und Schöne an diesem Buch ist der Einblick, den man gewinnt, wie damals die ‚Familie' der Atomphysiker zusammenarbeitete, ihr so oft räumliches und zeitliches Zusammensein, ihre Kontiguität der Gedanken. So tauchen häufig sehr persönliche Erinnerungen an Zeitgenossen auf, die zur gleichen Ebene der hochkarätigen Atomphysiker gehören und gehörten. Zum anderen ist es an vielen Stellen überraschend und ungewohnt, dem Menschen *Heisenberg* so intim zu begegnen: Er fährt Ski in den Alpen, diskutiert hart mit Kollegen über Atomphysik, Philosophie und Politik, seine Stellung während der finsteren Jahre des Faschismus in Deutschland wird aufgehellt, auch seine Erlebnisse im Rahmen des Uran-Vereins. Daß sein Alterswerk, die einheitliche Feldtheorie, deren zentrale Aussage unter dem Namen Weltformel bekannt geworden ist, gegen Ende der Biographie nur knapp gestreift wird, mag wohl daran liegen, daß gerade diese Arbeiten die neue Forschergeneration nicht mehr erreichten. Die Quarks der Quantenchromodynamik haben zu heftigen Auseinandersetzungen geführt, und *Heisenberg* mußte einräumen, daß es Quarks tatsächlich geben könne; ob ein Fehler oder ein neuer Weg, beides ist unlösbar mit der Natur menschlichen Wissens und Forschens verbunden.

Es ist ein Buch, das die ältere und jüngere Generation der Naturwissenschaftler gleichermaßen für lesenswert halten dürfte: Die Jüngeren erfahren etwas darüber, wie es damals war und wie damals Schulen wirkten und sich gegenseitig fruchtbar beeinflußten, und die Älteren können lesen, was sie selbst teilweise aus eigenem Erleben erfahren haben und wie sich diese Erfahrung aus der Sicht einer bestimmenden Forscherpersönlichkeit anhört. Für den Historiker der Naturwissenschaften ist es auch eine Quelle für Einblicke in die Atomphysiker-Generation.

H. Hübner

Bone and Bone Seeking Radionuklides
Physiology, Dosimetry and Effects
EULEP Symposium, Rotterdam, August 1980

V. Volf (Hrsg.). New York: Harwood Academic Publishers GmbH 1981. 153 S., div. Abb. u. Tab.; geb. $ 32.00; DFL 80,–

Die vorliegende Monographie basiert auf einem Symposium, organisiert von der EULEP (European late Effects Projekt Group). Der Inhalt umfaßt 3 Hauptthemen. Zunächst wird eine Übersicht über die Häufigkeit und Signifikanz von Knochentumoren sowie über Funktion, Aufbau und Stoffwechsel der Knochenzellen gegeben. Die neuesten Methoden zu Studium und Beschreibung der Aufnahme von Radionukliden im Knochen werden beschrieben. Die Mikroverteilung der Radionuklide stellt die Basis für die Dosimetrie dar. In weiteren Beiträgen wird die Verteilung von Plutonium-239 und die Ablagerung im Knochen beschrieben. Die Kinetik des Plutoniums, speziell der Übergang vom Blut in die Knochensubstanz und deren Rückverteilung über Osteoklasten und Makrophargen ins Blut und in die Knochensubstanz sowie die räumlichen und zeitlichen Verteilungen, werden anhand von autoradiographischen Untersuchungen an Mäusen untersucht. Die Alphadosimetrie wird mit neuen computergesteuerten quantitativen Methoden der Autoradiographie und Fluoreszenzphotographie sowie mit Mikrophotogrammen belegt.

Im 3. Abschnitt werden quantitative und biologische Aspekte der Entstehung von Knochentumoren durch Bestrahlung behandelt. Die onkogenetischen Effekte werden in Versuchen an der Maus mit kurzlebigen, knochenaffinen Alphastrahlern (Radium-242, Thallium-227) aufgezeigt.

Bestrahlungsqualität, Skelettdosis und zeitliche sowie räumliche Verteilung der Radionuklide werden diskutiert. In weiteren Versuchen mit dem kurzlebigen Beta-Strahler Lutetium-177 werden Vergleiche mit dem langlebigen Beta-Strahler ^{90}Sr vorgestellt. Beide Strahler zeigen die gleiche Wirkung, wobei bei protrahierter Gabe Osteosarkome häufiger auftreten.

In Mäuse- und Rattenversuchen wird der Einfluß biologischer Parameter, wie z.B. Geschlecht, Alter, Zeit der Exposition, hormonelles Gleichgewicht, Belastung und genetische Konstitutionen, beschrieben.

Die Publikation bringt wichtige Informationen für Strahlenphysiker, Strahlenbiologen und Strahlenärzte. Jedes Kapitel enthält ein informatives, umfangreiches Literaturverzeichnis für weitere Studien.

Die Forschungsergebnisse werden anhand von instruktiven Tabellen, schematischen Zeichnungen sowie photographischen Darstellungen von histologischen Bildern und Autoradiogrammen dargestellt, so daß auch für den Nichtspezialisten ein hoher Informationsgehalt besteht.

G. Schneider

Nuclear Medicine
Factors Influencing the Choice and Use
of Radionuclides in Diagnosis and Therapy

NCRP Report No. 70; National Council on Radiation Protection and Measurements, Bethesda 1982; 180 S.; Broschur $ 12.00

In der nuklearmedizinischen Praxis wird vom Arzt die Entscheidung verlangt, das für Diagnostik oder Therapie beste Radionuklid bzw. Radiopharmakon auszuwählen sowie das Meß-, Abbildungs- und Auswerteinstrumentarium festzulegen. Darüber hinaus sind Abschätzungen über die Strahlendosis notwendig. Dieser Report soll die Entscheidungen bei diesen Fragen erleichtern, indem er die derzeit vorliegenden Erfahrungen bzw. Daten, übersichtlich zusammengestellt, mitteilt.

Nach einem interessanten Rückblick auf die Geschichte der Anwendung von radioaktiv markierten Verbindungen in der Medizin wird ausführlich auf die wünschenswerten Daten für Radiopharmaka, die in Diagnostik oder Therapie eingesetzt werden sollen, eingegangen. Im einzelnen werden die optimalen Charakteristika für das Radionuklid, seine chemische Form sowie für die In-vivo-Verteilung besprochen, wobei auch Faktoren berücksichtigt werden, die Kinetik, Metabolismus und Verteilung beeinflussen. Nach einem kurzen Überblick über die in Reaktoren oder Beschleunigern produzierten Radionuklide werden u. a. zahlreiche Hinweise für analytische Untersuchungen gegeben, z.B. für die Identität, Reinheit und Pyrogenfreiheit der Präprapate.

Der zweite Hauptabschnitt des Buches befaßt sich mit den Geräten für nuklearmedizinische Untersuchungen. Hier erhält der Leser einen kompletten Überblick über das derzeit vorhandene Instrumentarium, wobei speziell Empfindlichkeit und Auflösung näher untersucht werden.

In dem Abschnitt über die absorbierte Strahlendosis nach Gabe von Radiopharmaka wird u. a. eine Reihe von Beispielen für Dosisabschätzungen gegeben, die für den behandelnden Arzt von großem Nutzen sein müssen. Das gleiche gilt für das Kapitel, in dem die zweckmäßigsten Einsatzgebiete von Radiopharmaka untersucht und an mehreren instruktiven Beispielen Figure-of-Merit-Abschätzungen angegeben werden.

Das zusammenfassende Schlußkapitel gibt praktische Richtlinien und Empfehlungen für nuklearmedizinische Untersuchungen.

Zwei Nachträge erhöhen den Wert des außerordentlich informativen Büchleins noch. Im ersten werden Richtlinien für die Bewertung von Radiopharmaka gegeben, die in erster Linie für die Entwickler neuer Präparate nützlich sind.

Der zweite Nachtrag enthält in Form einer umfangreichen Tabelle Angaben über die in den Geweben absorbierten Strahlendosen in mrd/μCi für die meisten derzeit angewendeten Radiopharmaka.

Obgleich sich das Buch vorwiegend auf Primärliteratur amerikanischer Autoren stützt und einige Ausführungen speziell auf US-amerikanische Verhältnisse zugeschnitten sind, kann doch sein Nutzen für alle in der Nuklearmedizin Tätigen gar nicht überschätzt werden. *H. Herzmann*

Physik der Hochschulanfänger

H. Wegener. Stuttgart: B. G. Teubner 1982. Bd. I: 258 S., 117 Abb., 7 Tab.; Bd. II: 272 S., 126 Abb., 4 Tab. (Teubner Studienbücher) kart. je Bd. DM 23,80

Die zweibändige „Physik für Hochschulanfänger" bringt die Grundlagen der Physik auf einem Niveau, das einer üblichen Experimentalphysikvorlesung entspricht Nach einer Einführung über Materie und Antimaterie, die Symmetrieüberlegungen enthält und das Interesse des Lesers wecken soll, wird die Punktmechanik, auch relativistisch, dargestellt. Relativitätsprinzip und *Lorentz*-Transformation werden besprochen, bevor noch die Grundgesetze der Mechanik behandelt worden sind. Bei der Integration der Bewegungsgleichungen wird gleich mit auf die numerische Integration derselben eingegangen. Die Erhaltungssätze der Mechanik und ihre Anwendungen, auch auf starre Körper, schließen sich an. Zur relativistischen Mechanik werden einige sehr schöne Anwendungsbeispiele vorgeführt. Der Abschnitt über die Gravitation enthält auch Interessantes über die Sternentwicklung. Die beschleunigten Koordinatensysteme schließen die Punktmechanik ab.

Relativ kurz werden einige mechanische Eigenschaften fester, flüssiger und gasförmiger Körper sowie Strömungen in Flüssigkeiten und Gasen dargestellt. Der erste Band endet mit Thermodynamik und Statistik, wobei im wesentlichen die Grundbegriffe eingeführt, 1. und 2. Hauptsatz und die kinetische Gastheorie behandelt werden.

Der zweite Band enthält vorwiegend die elektromagnetischen Erscheinungen. Zunächst werden Meßmöglichkeiten für Ladungen, Ströme und Felder wiedergegeben. Die *Maxwell*schen Gleichungen für Felder im Vakuum werden in integraler Schreibweise eingeführt. Die mathematischen und dann die physikalischen Eigenschaften der eingeführten Felder werden besprochen: Elektrostatik, Leiter im elektrischen Feld, Magnetostatik. Im Zusammenhang mit einfachen Stromkreisen wird auch auf die mikroskopischen Vorstellungen zur Leitfähigkeit von Elektrolyten, Metallen und Halbleitern eingegangen. Weitere Abschnitte sind *Faraday*sches Induktionsgesetz, Wechselstromkreise, elektromagnetische und mechanische Schwingungen und elektromagnetische Wellen. Ein kurzer Abschnitt behandelt Materie im elektromagnetischen Feld.

Im Kapitel über Wellenoptik und Phononen werden Beugungserscheinungen (auch *Bragg*-Reflexion), Brechung, Dispersion, *Doppler*-Effekt, Polarisation und Eigenschaften, die die Quantennatur des Lichtes zeigen, aufgegriffen.

Abschnitte über Materiewellen und Atomphysik, Atomkerne und Radioaktivität sowie über Quarks beschließen das Lehrbuch.

Die Darstellung des Stoffes ist klar, lebendig und ansprechend. Viele moderne Beispiele sind von besonderem Wert und richten den Blick des Lesers auf neuere Entwicklungen, insbesondere in der Elementarteilchenphysik. Bei den Anwendungen wird oft der Darstellung vorgegriffen, das ist bei der Bewegung in elektrischen und magnetischen Feldern besonders gut gelungen.

Weniger glücklich erscheint die Einführung der *Lorentz*-Transformation als experimentelles Ergebnis der Beobachtung der an einer Kugel reflektierten Lichtblitze von zwei Inertialsystemen aus. Das Beispiel, das den Widerspruch zwischen dem Begriff des starren Körpers und dem *Einstein*schen Relativitätsprinzip zeigen soll, ist wohl nicht zutreffend: Die beschriebene Zerstörung eines Gerätes könnte man z.B. über den *Doppler*-Effekt tatsächlich ohne Widerspruch zum Relativitätsprinzip erreichen; die Überlichtgeschwindigkeit der Signalausbreitung bei einer starren Stange ist im Buch nicht hervorgehoben.

Leider ist das SI nicht konsequent verwendet worden (cal, Abkürzungen: sec statt s, Cb statt C, abweichende Definition der Magnetisierung), und es sind einige unübliche Begriffe eingeführt worden (z.B. magnetische Erregung für H statt Magnetfeldstärke, magnetisches Feld für B statt magnetische Induktion oder magnetische Flußdichte).

Das insgesamt gut und modern angelegte Lehrbuch leidet etwas unter den erwähnten kleinen Unzulänglichkeiten. Es kann — mit geringen Einschränkungen — Studenten der Physik, Chemie, Mathematik und technischer Wissenschaften empfohlen werden; Lehrende der Physik werden es mit Vorteil zur Vorbereitung von Lehrveranstaltungen verwenden können. *A. Kühnel*

Repetitorium der Physik

F. K. Kneubühl, 2. überarb. u. erw. Aufl., Stuttgart; B. G. Teubner 1982. 544 S., 318 Abb., 67 Tab. (Teubner Studienbücher Physik) kart. DM 39,80

Die Punktmechanik steht naturgemäß am Anfang des Repetitoriums und wird mit einem Abschnitt über Maßeinheiten eingeleitet. Nach Kinematik und Grundlagen der Dynamik eines Massenpunktes wird der Gravitation besonderes Augenmerk gewidmet. Die Erhaltungssätze für Massenpunktsysteme und deren Anwendungen schließen sich an. Gleichgewicht und Stabilität werden mit Katastrophentheorie behandelt. Die Formalismen beschließen die klassische Punktmechanik.

Das zweite Kapitel beschäftigt sich mit der Relativität. Nach *Galilei*-Transformation und Bewegungen in bewegten Bezugssystemen werden die Grundlagen der Speziellen Relativitätstheorie dargestellt und deren wichtigste Konsequenzen für die Punktmechanik zusammengestellt.

Die Mechanik des starren Körpers umfaßt neben den Grundbegriffen und der Kinematik die Statik, den starren Rotator und den Kreisel. Die Mechanik der deformierbaren Medien wird ausführlich berücksichtigt. Der Begriff der Spannung, Statik und Kinematik von Flüssigkeiten und Gasen bereiten die Dynamik von idealen und viskosen Flüssigkeiten vor. Weitere Abschnitte befassen sich mit Potentialströmungen, Wirbeln, Überschallströmungen, Turbulenz und dynamischem Auftrieb. Das Kapitel über Elektrizität und Magnetismus enthält die Gebiete Elektrostatik, stationäre elektrische Ströme, Magnetismus und quasistationäre Ströme. Die Abschnitte über dielektrische Eigenschaften der Materie, elektrische Leiter und magnetische Eigenschaften der Materie behandeln neben phänomenologischen auch mikroskopische Aspekte.

Elektromagnetischen und mechanischen Schwingungen und Wellen ist das umfangreichste Kapitel vorbehalten. Die Optik, einschließlich geometrischer Optik, wie auch die Abstrahlung elektromagnetischen Wellen finden hier ihren Platz.

Quanten- und Wellenmechanik beginnen mit dem Welle-Teilchen-Dualismus. Es folgen die Grundbegriffe der Wellenmechanik, der lineare harmonische Oszillator, das Wasserstoffatom, ein Elektron im periodischen Potential und die Eigenschaften von Drehimpulsen (auch Spin) und magnetischen Momenten.

Das Kapitel über Thermodynamik bringt die drei Hauptsätze und die Wärmeleitung allgemein sowie ideales und *van-der-Waals*-Gas als Modellsubstanzen. Das kurze Kapitel über statistische Mechanik enthält *Boltzmann*-Statistik, kinetische Gastheorie und Quantenstatistik. Als Anwendungen findet man das *Einstein*sche Modell der spezifischen Wärme, das Elektronengas in Metallen und die Wärmestrahlung.

Das letzte Kapitel befaßt sich mit Atomkernen und Elementarteilchen: Aufbau der Atomkerne, Radioaktivität, Kernreaktionen und Elementarteilchen.

Fünf Anhänge führen Lehrbuchliteratur, die physikalischen Einheiten einschließlich der Umrechnungen zwischen den verschiedenen Maßsystemen, physikalische Konstanten und Tabellen, mathematische Tabellen und Fachwörter der Physik in Englisch, Deutsch und Französisch an. Ein detailliertes Inhaltsverzeichnis und ein Sachverzeichnis erleichtern dem Leser das Auffinden eines Sachgebietes.

Das Repetitorium der Physik ist sehr zu empfehlen für diejenigen, die sich mit den darin enthaltenen Gebieten bereits befaßt haben und ihr Wissen auffrischen oder ergänzen wollen, vor einer Prüfung oder für die eigene Arbeit in Lehre oder Forschung. Es ist keine reine Formelsammlung: Wo es in der gebotenen Kürze möglich ist, wird auf Erläuterungen und Herleitungen nicht verzichtet. Eine Vielzahl von Tabellen vermittelt eine Vorstellung von den in der Realität anzutreffenden Werten physikalischer Größen. Viele gelungene Abbildungen fördern das Verständnis. Besonders erfreulich ist die durchgängige Benutzung der SI-Einheiten. Das Niveau entspricht einem üblichen Grundkurs der Physik.

A. Kühnel

Physikalische Chemie
Teil I: Atome, Moleküle, Kerne

G. M. Barrow: Bearb.: G. W. Herzog, 5. neubearb. Aufl.; Wien: Bohmann-Verlag; Braunschweig — Wiesbaden: Fr. Vieweg & Sohn, 1982, 374 S., 168 Abb., 44 Tab., Paperback DM 26,80

Im Jahre 1660 erschien in Paris ein zweibändiges Werk von *Nicolas Lefèvre* (La Fabure) mit dem Titel „Traité de la Chymie". Dort wird den Ärzten und Apothekern zur Vermeidung von Irrtürmern bei der Bereitung von Medikamenten empfohlen, Zuflucht in die physikalische Chemie „... avoir recourse à la Physique Chymique ..." zu nehmen. Rund 100 Jahre später ermahnt *Torben Bergman*, ein Chemie-Professor in Uppsala, seine Schüler, daß man auch ein guter Physiker sein müsse, um ein guter Chemiker werden zu können, und nochmals 100 Jahre später fordert *Robert Bunsen* dem Chemiker von Geltung auch gründliche Kenntnisse in Physik und den mathematischen Hilfsmitteln ab. Das gesamte Gebiet, das so häufig direkt oder indirekt gefordert wurde, ist seit vielen Jahren ein selbständiger Zweig der Chemie — die physikalische Chemie —, und man kann heute fragen, welche Forderungen an Kenntnissen dem angehenden Chemiker, Verfahrenstechniker und auch dem Physiker abverlangt werden, um im Beruf seinen Mann zu stehen, sei es in Forschung oder Lehre, sei es in Labor oder Betrieb.

Im Jahre 1961 erschien die erste amerikanische Auflage der Physical Chemistry von *Gordon M. Barrow* bei McGraw-Hill in New York. Seither sind mehrere Auflagen und Übersetzungen erschienen; die 5. neubearbeitete deutsche Auflage liegt hier vor. Offensichtlich hat sich diese Art der Darstellung des Stoffes und der Wissensvermittlung mehrfach überzeugend bewährt, was man schon beim ersten Anblättern des Buches bemerken kann. Der 1. Teil ist der Atomistik gewidmet, beginnend mit den Gasgesetzen. Übrigens werden die gemessenen Zahlenwerte von *Robert Boyle* wiedergegeben, die ihn zu seinem berühmten Gesetz führten (1660—64). Einen verhältnismäßig breiten Raum nimmt die Anwendung mathematischer Methoden auf diesem Gebiet ein: Molekülsymmetrie und Gruppentheorie, *Hermite*sche-, *Legendre*sche- und *Laguerre*sche Polynome, die in der Quantenmechanik bedeutsam sind. Inwieweit allerdings auf einer halben Seite die *Fourier*transformation noch durchschaubar gemacht werden kann, sei dahingestellt.

Einige Bemerkungen aus der Sicht der Isotopenchemie, Atom- und Kernphysik seien angeführt: Die Massenzahlen der Nuklide werden uneinheitlich angewendet, wobei es keine Frage der Einhaltung einer JUPAC-Empfehlung ist, dies links oben am Elementsymbol vorzunehmen. So kommt es beispielsweise auf S. 240, Aufgabe 3, zu der Wiedergabe von $^{39}KCl^{35}$. Die Isotopentabelle auf S. 354 ist zwar auf das Nuklid ^{12}C bezogen, verwendet aber noch den Begriff des chemischen Atomgewichtes, der eigentlich hinfällig ist. Diese Tabelle sollte außerdem mehr als nur auf 4 Stellen das Atomgewicht enthalten (JUPAC 1971), um einerseits die natürlichen Schwankungen der Atomgewichte einiger Elemente deutlich zu machen, andererseits den Stand in der Meßtechnik für einzelne Elemente wiederzugeben. Die auf S. 223/24 vorgestellte Photoelektronenspektroskopie macht den Unterschied zwischen einer UV-Photonenspektroskpie (UPS) und einer Röntgen-Photonenspektroskopie (XPS) nicht deutlich, noch dazu, da ESCA nur als Literaturhinweis zur Abbildung 6.25 erwähnt wird. Physikalische Chemie heute ist auch die Wiedergabe der Grundlagen einer Reihe von modernen instrumentellen Analyseverfahren (PIXE, NAA, AAS u. a. m.). Gleiches gilt auch für die Grundlagen der Laser-Technik, die neben ihrer Verwendung in der analytischen Chemie auch für Isotopentrennverfahren zunehmend an Bedeutung gewinnt und nicht ohne weiteres verstanden werden kann. Die Elementarteilchen der Atomkerne und die Erwähnung der Quarks (S. 319) sollten aktualisiert (J/ψ-Teilchen 1974) und leichter durchschaubar gemacht werden. Die Isotopieverschiebungen in den Schwingungsspektren der Moleküle sind nur in einigen Aufgaben (S. 240) enthalten; da sie den einzigen Zugang zum Verständnis der unterschiedlichen Verhaltensweisen isotoper Moleküle bilden, sollte ihnen auch mehr Raum bereitgestellt werden. Die Tabelle 3.1 auf S. 34 ist die Tabelle 2.1.

Das Lehrbuch ist eine systemvolle und konzentrierte Wiedergabe des angehäuften Wissens auf dem Gebiet der physikalischen Chemie und wird in einem hohen Maße den gestellten Forderungen nach gründlichen Kenntnissen in der Physik und in den mathematischen Methoden gerecht, wie sie Chemiker zur Beantwortung der auftauchenden Fragen nach dem Warum bestimmter Eigenschaften und dem Verhalten von Stoffen gegenüber äußeren Einflüssen benötigen.

H. Hübner

Einführung in die Teilchenoptik

J. Großer. Stuttgart: B. G. Teubner 1983. 155 S., 71 Abb. (Teubner Studienbücher) Kart. DM 21,80

Die wichtigsten Forschungsergebnisse auf dem Gebiet der Elektronen- und Ionenoptik wurden in den Jahren 1930 bis 1960 erzielt und auf ihrer Grundlage Geräte wie das Elektronenmikroskop, das hochauflösende Massenspektrometer und die ersten Beschleuniger entwickelt. Heute ist dieses Teilgebiet der Physik kein aktuelles Forschungsgebiet mehr. Dennoch kommt der experimentell arbeitende Physiker, der sich mit Elektronen- oder Ionenstrahlen beschäftigt, nicht umhin, Hilfsmittel aus der geometrischen Teilchenoptik anzuwenden.

In seinem Buch „Einführung in die Teilchenoptik" gibt der Verfasser einen breiten Überblick, wobei der Schwerpunkt auf der Charakterisierung von Bauelementen teilchenoptischer Geräte wie z. B. Linsen und Prismen liegt. Er beschreibt ohne großen mathematischen Aufwand ihre Eigenschaften und Wirkprinzipien sowie ihr Zusammenwirken in teilchenoptischen Geräten.

Ausgehend von den Grundgleichungen für die Bewegung elektrisch geladener Teilchen unter Einfluß elektrischer und magnetischer Felder und den Ähnlichkeitsgesetzen für deren Bahnen im entsprechenden Feld werden Linsen und ihre Kombinationsmöglichkeiten beschrieben. Auftretende Abbildungsfehler, die Besonderheiten der Optik im relativistischen Bereich, Transfermatrizen für Linsenkombinationen und einige Anwendungsbeispiele runden den Inhalt der ersten beiden Kapitel ab.

In einem weiteren Kapitel beschäftigt sich der Autor mit Prismen und Spektrometern. Zylinderkondensator- und Sektorfeldberechnungen werden in einfacher Weise dargestellt, wobei auch Randeffekte und auftretende Abbildungsfehler diskutiert werden. Mehrfachfokussierende Anordnungen und andere Spektrometertypen wie z. B. Wienfilter, Laufzeit- und Hochfrequenzspektrometer und Quadrupolmassenfilter werden erläutert.

Die Raumladungseffekte, die in Strahlen geladener Teilchen bei entsprechender Teilchendichte auftreten, deren Einflüsse auf die Strahlverbreitung und Energieverminderung und die Möglichkeiten der Beeinflussung dieser Effekte durch spezielle Anordnungen schließen die Behandlung der Optik geladener Teilchen ab.

Im letzten kurzen Kapitel ist die Optik mit neutralen Atomen und Molekülen beschrieben. Linearer und quadratischer Starkeffekt, *Zeeman*effekt und quadratischer *Zeeman*effekt werden näher erläutert. Auf eine Darstellung von numerischen Lösungsmethoden für teilchenoptische Probleme verzichtet der Autor ebenso wie auf die exakte Berechnung von Abbildungsfehlern und Abbildungen höherer Qualität.

Das Taschenbuch wendet sich an Wissenschaftler und Studenten, die sich einen Überblick über das Gebiet der Teilchenoptik verschaffen wollen. Für einfache Abschätzungen reichen die in diesem Buch aufgezeigten Lösungswege aus. Auch für das Verständnis weiterführender Literatur schafft sein Studium gute Grundlagen.

H. Neumann

Die gaschromatographisch-massenspektrometrische Spurenanalyse organischer Substanzen in der Luft

W. A. Issidorow u. I. G. Senkewitsch. Moskau: Isdwo-Chimija. 1982. 136 S., 19 Abb., 26 Tab., 280 Lit. (in Russisch) kart. M 7,50

Die auf dem Gebiet der Umweltanalytik international bekannten Autoren der Chemischen Fakultät der Leningrader Universität wenden sich mit diesem Büchlein einer hochaktuellen Problematik zu, der weltweit zunehmende Aufmerksamkeit gewidmet wird.

Im ersten Kapitel werden die chemische Zusammensetzung der Atmosphäre und das Ausmaß des biogenen und anthropogenen Eintrages organischer Verbindungen in dieselbe vorgestellt. Weiterhin werden die Emissionsquellen organischer Verbindungen in Großstädten (Verkehr, Industrie, Kommunalwirtschaft) diskutiert und in mehreren Tabellen qualitative und quantitative Werte einer Vielzahl nachgewiesener Verbindungen (Kohlenwasserstoffe, sauerstoff-, stickstoff-, schwefel- oder halogenhaltige Substanzen) angegeben. Neben Werten von westeuropäischen und amerikanischen Großstädten findet man auch entsprechende Konzentrationsangaben sowjetischer Städte (Leningrad, Baku, Tbilissi, Taschkent, Murmansk u. a.).

Im zweiten Kapitel werden die Methoden zur Anreicherung der Spuren organischer Verbindungen aus der Luft durch Ausfrieren oder durch Adsorption an anorganischen oder organischen Materialien im Gradientenrohr vorgestellt und aus praktischer sowie theoretischer Sicht diskutiert.

Die thermische Desorption der angereicherten Proben in die gaschromatographische Apparatur, die Trennung und Identifizierung anhand von Retentionswerten wird im Kapitel 3 beschrieben.

Im vierten, dem umfangreichsten Kapitel wird die Identifizierung der organischen Spurenkomponenten mit Hilfe der GC-MS-Kombination abgehandelt, wobei besonderer Wert auf die Anwendung verschiedener Algorithmen zum Vergleich von Massenspektren mit Literaturwerten gelegt wird.

Die Angabe der Massenspektren von 235 umweltrelevanten organischen Verbindungen als Anhang ergänzt die Informationen. Dieses Buch schließt eine Lücke auf dem Gebiet der Umweltanalytik organischer Verbindungen. Organische Spurenanalytiker, Lufthygieniker, Umweltanalytiker, Toxikologen u. a. werden es mit Gewinn lesen, ohne auf das Studium von Lehrbüchern der Chromatographie und Massenspektrometrie verzichten zu können.

R. Herzschuh

Moderne Probleme der atmosphärischen Diffusion und der Verschmutzung der Atmosphäre

M. E. Berljand; in deutscher Sprache hrsg. von H. Ihlenfeld. Berlin: Akademie-Verlag 1982. 435 S., 81 Abb., 43 Tab.; Leinen M 56,—

Eines der Medien, in welches moderne Industrien ihre Schadstoffe abgeben, ist die Atmosphäre. In ihr finden neben dem Transport der Schadstoffe von der Emissionsquelle zu mehr oder weniger entfernten Orten vor allem auch chemische Umwandlungen statt. Ein Hauptprozeß der Transportvorgänge in der Atmosphäre ist die Diffusion, die hier zum Gegenstand einer ausführlichen Darstellung gemacht wurde und den gegenwärtigen Stand auf diesem Gebiet zusammenzufassen versucht. Der Autor — ein namhafter sowjetischer Wissenschaftler — untergliedert die atmosphärische Diffusion in insgesamt 14 Kapitel unterschiedlicher Länge, wobei vor allem dem mathematischen Apparat zur Behandlung des Themas eine beachtenswerte Breite eingeräumt wird.

Einer solchen Zielstellung kann nur zugestimmt werden, und eine solche Darstellung wird in weiten Kreisen von Umweltforschern und Aerochemikern begrüßt werden, denn nur wenig zusammenfassendes Material dieser Art ist im Schrifttum zu finden. Einige Mängel sind aber nicht zu übersehen. Den häufig wechselnden Abkürzungen hätte man durch ein umfangreiches Abkürzungsverzeichnis mit Seitenangaben begegnen können, da beispielsweise allein für die Konzentration eines Stoffes in der Atmosphäre unterschiedliche Abkürzungen ($\bar{q}$, s, c_L, c_M, q_M, q) gewählt werden. Das Inhaltsverzeichnis bedarf auch einiger notwendiger Erweiterungen: *Karmán*-Konstante, S. 193 u. 228; Tracer S. 259, SF_6 S. 259, ^{41}Ar S. 255 u. 398 u. a. m. Zur Fotosynthese wäre noch die S. 390 zu ergänzen, dafür aber der Hinweis auf die S. 386 zu streichen, da in der Stratosphäre zwar eine Vielzahl fotochemischer Reaktionen ablaufen, die aber nicht zur biochemischen Fotosynthese zu zählen sind — übrigens fotochemisch oder photochemisch (S. 152 u. 153). Das PAN auf S. 153 ist Peroxy . . . und nicht Peroxiazetylnitrat, und der Hinweis auf die atmosphärischen Stabilitätsklassen S. 254 soll Abschnitt 7.7 und nicht 7.6. bedeuten. Den chemischen Umwandlungen in der Atmosphäre (Troposphäre und Stratosphäre) wurde nur wenig Beachtung geschenkt. Beispielsweise wird von der Umwandlung von SO_2 in H_2SO_4 nur angenommen, daß eine Reaktion von atomarem Sauerstoff mit SO_2 stattfindet. In Rauchfahnen dürfte aber eher eine Bildung von $SO_2 + H_2O \rightarrow H_2SO_3$ stattfinden, das sehr leicht zu H_2SO_4 autoxydiert wird. Gar nichts wird über Isotopenvariationen des Schwefels in verschiedenen Schwefelverbindungen der Atmosphäre (SO_2, SO_4^{--}) gesagt, denn beispielsweise wurden die *Keystone*-Experimente S. 257 auch unter diesem Aspekt ausgeführt.

Die Literatur wurde mit insgesamt 655 Zitaten erfaßt, allerdings mit den letzten Zitaten aus dem Jahre 1979. Auf die Jahre 1976—1979 entfallen nur 12,7%, das Maximum liegt bei 1971 mit 13,3%. Dies dürfte auf die unvermeidliche Zeitverzögerung durch die Übersetzung zurückzuführen sein. Manche Zitate sind ungenau und unvollständig: J. geophysic. Res. wird immer ohne Jahrgang zitiert und Atmosph. Environment mit ungenauen Angaben der Jahrgänge. Die Probleme des globalen CO_2-Anstiegs oder einer eventuellen Ozonschildzerstörung in der Stratosphäre durch anthropogene Einflüsse sind nur bedingt als aktuell dargestellt einzuordnen, da gerade in den letzten Jahren ein umfangreiches Schrifttum darüber erschienen ist. Ähnliches gilt auch von den Abgaben schwacher Aktivitäten aus der kerntechnischen Industrie (T, ^{14}C, ^{85}Kr u. a.) an die Atmosphäre.

Der Vorgang der Diffusion von Schad- und Spurenstoffen in der Atmosphäre ist nur ein Teilprozeß neben den chemischen Umwandlungen, Auswaschungen (saurer Regen) und Ausfällungen (fall-out). Gerade beim Problem des Langstreckentransportes können die anderen Teilprozesse nicht vernachlässigt werden. Die Monographie ist zwar eine wertvolle Zusammenstellung von Fakten diffuser Vorgänge in der Atmosphäre, bedarf aber einer stärkeren Einbeziehung anderer Teilprozesse und einer notwendigen Aktualisierung.

H. Hübner

Unsere Erde wird überleben
GAIA — Eine optimistische Ökologie

J. E. Lovelock München — Zürich: R. Piper & Co. 1982, 223 S., 7 Abb., 3 Tab.; geb. DM 28,—

Bei den Griechen war es Gaia, bei den Römern Tellus, die seit ältester Zeit verehrte Göttin Erde, die alles Leben aus ihrem Schoße gebiert, an ihrer Brust trägt und nährt. Auch bei Homer wird diese glorreiche Göttin bei Eidschwüren neben Zeus, Helios und Unterwelt angerufen und erhält als Opfer ein schwarzes Lamm. Und gerade diese Erdgöttin steht heute als Symbol für die These von der gleichen Erde als eines sich selbstregulierenden Ökosystems, das sich in Jahrmillionen entwickelt und sich der ständig zunehmenden Eingriffe des Menschen zu erwehren hat.

Der Autor, ein in England lebender unabhängiger Naturwissenschaftler, versucht ein Bild der Erde zu entwerfen, wie sie sich als ein aktives Regelglied in ihrer Bedrohung durch die vielseitigen anthropogenen Belastungen verhält, wobei der Biosphäre der Hauptanteil dieses Regelmechanismus zukommt. Daß so viele Menschen heute die Erde als ein Ganzes betrachten und auch betrachten wollen, dürfte nicht zuletzt auf den weitbekannten Anblick unserer Erde aus dem Weltraum zurückzuführen sein: unsere Erde, der blaue Planet, das einzige Weltraumschiff, das in einem Umkreis von vielen Lichtjahren eine Biosphäre hat und von vernunftbegabten Wesen bewohnt ist.

Eines bleibt unklar oder wird zumindest nicht mit der Deutlichkeit ausgesprochen, wie man es in einer ohne Zahlen und Modelle gewählten Darstellung doch sagen sollte: Warum oder wann ist denn Gaia, die Erde, durch Umweltbelastungen bedroht? Doch wohl immer dann, wenn im Kreislauf der Stoffe in der Natur dieser Kreis eben nicht geschlossen ist, auch nicht durch integrierende Beteiligung der Biosphäre, der Mikroorganismen; wenn anthropogene Stoffe erzeugt werden, die nicht wieder oder nur mit großen Zeitkonstanten abgebaut und dem Kreislauf aufs Neue zugeführt werden. Es ist keine Frage der Einstellung, ob das Spraydosenproblem als hysterisch überhöht eingestuft wird oder ob N_2O als Ozonschildzerstörer wirkt oder nicht. Beide Spurengase werden in der Troposphäre nicht (und das N_2O auch in der Biosphäre nicht genügend schnell) abgebaut und gelangen eben in die Stratosphäre, wo erst Abbaumechanismen einsetzen können. Wenn eine Mülldeponie schneller wächst als ihr natürlicher Abbauvorgang, dann wird eben ein Problem daraus. Und wenn langlebige Transurane anfallen, die keinerlei Eigenschaften als Bioelemente haben, dann kann nur eine sorgsame Lagerung mit langen Wartezeiten ihren Abbau herbeiführen: Hier existiert kein Kreislauf.

Manches ist wohl durch die Übersetzung ins Deutsche ungenau wiedergegeben worden: So ist wohl *Lynn Margulis*, die bekannte mikrobiologische Ökologin aus Boston, eine Frau und nicht wie auf S. 11 ein Mann. Es gibt nur ein Enzym Carboanhydrase, das den Übergang von CO_2 in HCO_3^- und umgekehrt katalysiert, und nicht eine Transfer-Anhydrase S. 47 oder Karbonhydrase S.120. Es gibt nur negativ geladene Aluminat- und Silikat-Ionen, aber keine Al- und Si-Ionen, S. 134. Die Halogenkohlenwasserstoffe S. 149 finden auch andere Bezeichnungen: Halogenkarbonate S. 149, Chlor-Fluor-Kohlenstoffe S. 164/165, Karbontetrachlorid S. 149/150. In Stromatolithen kommt wohl Siliziumdioxid, aber nicht Si S. 50 vor, und noch stärker als zu Fe(II) reduziertes Eisen ist dann wohl freies Eisen S. 51. Von dem vielzitierten Schwefeldioxid wird nur gesagt, daß es durch seine zerstörende Wirkung auf Pilze zu einem besseren Gedeihen der Rosen in London führt, die im Vergleich dazu in ländlichen Gegenden weniger gut vorankommen – eine fast idyllische Präsentation des SO_2-Problems auf S. 157.

Auf weiten Strecken liest sich das Buch gut und aufklärend. Ob nur auf dem Lande lebende Wissenschaftler ein echtes Verhältnis zur Natur aufbringen können, wie es auf S. 193 angedeutet wird, kann wohl nicht als verbindlich angenommen werden. Solche und ähnliche Bücher sollen und können ein Schritt hin zur Natur sein, wie tief die Verehrung der Erdgöttin danach auch sein wird, denn wir können uns keinen anderen Planeten aussuchen.

H. Hübner

Praktische Mathematik

F. Stummel u. K. Hainer. 2. überarb. u. erw. Aufl.; Stuttgart: B. G. Teubner 1982 · 367 S., 16 Abb., 63 Übungsaufgaben mit Ergebnissen (Teubner Studienbücher: Mathematik) kart. DM 36,–

Vor allem im naturwissenschaftlich-technischen Bereich wird die Suche nach exakten Lösungen zunehmend durch rechnergestützte Näherungsberechnungen (unter weitgehender Nutzung von Standardprogrammen) abgelöst. Dafür benutzte Verfahren, die mit Namen wie *Newton*, *Euler*, *Gauß*, *Jacobi* belegt sind, zeigen, daß dieses sich immer noch stetig entwickelnde mathematische „Werkzeugreservoir" häufig schon seit vielen Generationen bereitliegt. Immer effektivere Bearbeitungsmöglichkeiten führen jedoch im Zeitalter des programmierbaren Taschenrechners zu einer rasanten Zunahme der Anwendungshäufigkeit von oftmals adaptierten Methoden der praktischen Mathematik insgesamt.

Das vorliegende Buch – hervorgegangen aus Vorlesungen und mathematischen Praktika – gibt einen modernen, wenn auch oft recht knappen Überblick über die grundlegenden Verfahren zur Funktions- und Nullstellenberechnung, Interpolation, numerischen Infinitesimalrechnung, Lösung algebraischer Gleichungssysteme und Eigenwertaufgaben bei Matrizen sowie numerischen Behandlung von Anfangswertaufgaben gewöhnlicher Differentialgleichungen. Die zweite Auflage ist durch ein Kapitel über die Fehleranalyse numerischer Algorithmen ergänzt worden, das speziell auf die Arbeit mit digitalen Rechenmaschinen zugeschnitten ist. Während auf die Intervall-Mathematik nicht eingegangen wird, kann die entwickelte Vorwärtsanalyse als wesentliche Bereicherung angesehen werden, die sowohl Datenfehler als auch gleitpunktarithmetische Rundungsfehler berücksichtigt.

Der Wert des Buches wird durch eine klare Gliederung, die 10% des Gesamtumfanges ausmachenden Übungsaufgaben und durch die zu jedem Abschnitt aufgeführten insgesamt etwa 100 in deutscher oder englischer Sprache herausgegebenen Literaturquellen gesteigert. Obwohl im Vergleich mit anderen Standardwerken mehr als Lehrbuch denn als Anleitung zum praktischen Handeln gedacht, sollte für eine weitere Auflage erwogen werden, dem Wunsch nach programmierten Algorithmen innerhalb des Buches zu entsprechen.

E. Serick

ISOTOPENPRAXIS

Volume 19 1983 Number 10

Contents

Articles

Letters to the Editor

Inhalt

Originalarbeiten

Kurze Originalmitteilungen

Isotopenpraxis is indexed in Current Contents/PC&ES, Science Citation Index, and INIS Atomindex

FSC
www.fsc.org
MIX
Papier | Fördert gute Waldnutzung
FSC® C083411

Kolibri 360 GmbH
Ferdinand-Jühlke-Straße 7
99095 Erfurt, Deutschland
produktsicherheit@kolibri360.de